Teilhard de Chardin:

The Search For The Light In Evolution

Bernard J. Fleury, Ed.D.

Teilhard de Chardin The Search For The Light In Evolution

Copyright © 2018 Bernard J. Fleury, Ed.D. All rights reserved.

Kurt E. Miller, Technology Consultant.

No part of this e-book may be reproduced, stored in a retrieval system, or transmitted by any means without written permission of the author.

Printed by Create Space, an Amazon.com Company.

ISBN-13: 978-1729577240

ISBN-10: 1729577245

Website: www.intolifebylight.com
Printed in the United States of America

Bernard J. Fleury Ed.D.

Dedication

To the Sulpician Fathers who first

Introduced me to the concept of

Evolution at St. Charles College,

Catonsville, Maryland in 1949 -1951.

Teilhard de Chardin The Search For The Light In Evolution

TABLE OF CONTENTS

Reviews 7

Description 10

Purpose Statement 12

Preface 13

Teilhard's Special Words and Phrases in Simple Common Use Terms 15

Pierre Teilhard de Chardin: The Man Who Christianized Purpose Driven Evolution 17

Prologue: Seeing 20

Teilhard de Chardin and Arthur Zajonc Synthesizers 25

The First Basic Point of Teilhard's Process of Evolution 27

The Body (Mass) Called Light 29

Was Rene` Descartes Dream a Vision or an Illusion? 30

Wave Theory of Light 31

The Light of Electricity: A Way of Seeing 34

Light and Magnetism 37

The Second Basic Point of Teilhard's Process of Evolution ... 39

Introduction to Teilhard's Purpose Driven Evolution 40

In the Beginning Was The Fire 41

Light: Foundational Energy of the Universe 42

Part I: Prelife 42

Part II: Life 49

Part III: Thought 54

Part IV: Superlife 62

Teilhard de Chardin: Essential Points On Man, The Creature Who Mirrors The Light 73

Summary 77

Bibliography 81

Epilogue 83

Author:
- Meet Bernard Video at http://www.intolifebylight.com
- LinkedIn Profile at
  www.linkedin.com/pub/bernard-fleury/41/115/a44/
- Amazon Profile at
  www.amazon.com/author/bernardfleury

Please note that a Table of Contents for Each of the five e-Books/Audio Books in the Series will be on website http://www.intolifebylight.com as each e-Book/Audio Book is published.

How Jesus Christ Leads Us To The Kingdom Of Heaven (Called Into Life By The Light Series Book 1) e-Book and Audio Book Published.

How *The Mind-Body-Spirit Connection in the Medicine of Light* affects medical practice and technology that promotes my well being. (Called Into Life By The Light Series Book 2) e-Book and Audio Book Published.

What does *Teilhard de Chardin: The Search For The Light In Evolution* reveal about my place as a human person in the development of the Earth? (Called Into Life By The Light Series Book 3) e-Book and Audio Book.

How does *Near Death Experience: Out Of The Darkness Into The Light* reflect what happens to persons who have had a profound near-death experience that may really happen to *me*

Teilhard de Chardin The Search For The Light In Evolution

when I die? (Called Into Life By The Light Series Book 4) e-Book and Audio Book.

How does *The Unfailing Light: Remedy For Alienation, Loneliness And Despair* actually help me deal with Alienation, Loneliness, and Despair? (Called Into Life By The Light Series Book 5) e-Book and Audio Book.

Come along with me as we explore the answer to the question what does

Teilhard de Chardin: The Search For The Light In Evolution

reveal about my place as a human person in the development of the Earth?

Bernard J. Fleury Ed.D.

Reviews

Teilhard de Chardin: The Search For The Light In Evolution is book three in the Called into Life by the Light Series by Bernard J Fleury, a slim book that will appeal to religious thinkers and philosophers, an essay inspired by the work of Teilhard de Chardin. What if there was something else in the beginning, something other than "the word"? What if there was light at the heart of creation, a fire that sets everything in motion?

In this essay, the author follows the thoughts of de Chardin and looks past the controversy around the origin of the world to appreciate the place of humanity in evolution. He explores Teilhard de Chardin's concept of purpose-driven evolution and maintains that from the moment humanity comes into the history of the universe, that history begins to have a purpose, a perspective, because man is part of the evolution that understood — the part of it that could see. The author will demonstrate that even evolution isn't a product of chance, but an unfolding of the light, of the will of God, and in this evolutionary trajectory humanity defines its place in the economy of God's saving plan.

Bernard J Fleury's book offers a fresh interpretation of the "phenomenon" of man, and where he fits into God's plan of creation. In this book, the author explores the concept of "seeing" properly in order to understand creation and makes apt references to other thinkers and theologians, featuring the ancients as well as modern thinkers — Leonard Euler, Robert Boyle, Thomas Young, Augustin-Jean Fresnel, Francois Arago, Michael Faraday, James Clerk Maxwell, and many others. Teilhard de Chardin: The Search For The Light In Evolution is a book that gives a new meaning to light as a creative property of God. This essay is well-researched and presented with the clarity of one who knows what they are talking about. It's as informative as it is inspiring.

Reviewer: Romuald Dzemo, Readers' Favorite, Rating *****

Teilhard de Chardin The Search For The Light In Evolution

Teilhard de Chardin: The Search For The Light In Evolution is a work of non-fiction by author Bernard J Fleury. The book forms Volume three in the Called into Life by the Light Series, and focuses mainly on philosophical issues posited by religion and belief. In his work, Fleury summarizes an accessible formation of the principles of Teilhard de Chardin, whose notion of Purpose Driven Evolution attempts to gel Darwinian thinking and scientific theory with the spiritual, overarching concept of God, the Creator and the Light. Fleury represents Chardin's philosophy through this metaphor of evolutionary stages, and presents an uplifting message of hope and unity across the universe.

I am not a spiritual person by nature, though I have a keen interest in philosophy, but Bernard J Fleury presented a concept which intrigued me from the first page. Fleury's retelling of Chardin's teachings is concise yet emotive, and it's very clear that the author has a true passion for their subject. Having said that, nowhere in the book did I find a message too pushy or judgmental. Teilhard de Chardin: The Search For The Light In Evolution is a well-balanced philosophical take, in which Fleury explains Chardin's viewpoint whilst also adding examples of his own from modern scientific thinking to exemplify his ideas. If you're looking for a thoughtful exploration of Purpose Driven Evolution, or indeed just a brand new way of thinking to stretch your brain muscles, you'll get a lot out of Teilhard de Chardin. Overall, a fascinating and well-compiled read.

Reviewer: K.C. Finn, Readers' Favorite, Rating *****

Teilhard de Chardin: The Search For The Light In Evolution by Bernard J Fleury is the third installment of the Called into Life by the Light five-book series, preceded by How Jesus Christ Leads Us to the Kingdom of Heaven and The Mind-Body-Spirit Connection in the Medicine of Light. Pierre Teilhard de Chardin was a Jesuit priest, geologist, and paleontologist in mid-twentieth century France, who traveled the world in the pursuit of answers that might help bridge the gap often associated between faith and science. Fleury has taken his work and composed it to further expand on Teilhard de Chardin's theories as they apply toward a balance between the concept of evolution, humanity, and the development of the earth, and the teachings of the Bible.

Teilhard de Chardin by Bernard J Fleury is an excellent resource for scholars, students, and scientists, as well as Christians who find themselves questioning either the aspects of scientific theory, the scripture of the Bible, or both. It's written concisely in an academic manner by Fleury, which is fitting given that he teaches as a college professor. In this book, Fleury digs even deeper, addressing the literary work and hypotheses of multiple intellectuals, both past and present, to present a compelling case. "Light is our origin, our mainstay, and our destiny. It is what forms and informs us. We are called into life by The Light, we live in The Light, and our destiny is The Light." I would recommend this book to any and all who are, as we all should be, open to alternate ideas based on a convergence of ideology and academia.

Reviewer: Jamie Michele, Readers' Favorite, Rating *****

Description

Achieving a knowledge of Purpose Driven Evolution, an understanding of the importance of the human person, **me**, in the development of the Earth. **What if** "seeing correctly" is the only way for me to see my real place in the development of the Earth. **What if** in order to see myself completely and correctly I have to see myself as part of humanity, humanity as part of life, and life as part of the universe.

As the title *Teilhard de Chardin: The Search For The Light In Evolution* suggests, this e-book is a presentation of Teilhard de Chardin's concept of Purpose Driven evolution.

As the e-Book cover illustration shows, "In the beginning was The Fire – The Big Bang" – the physical world thrust into existence and motion by The Light. All of creation is Purpose Driven.

Darwin's famous thesis that evolution proceeds by chance, groping, and the survival of the fittest could be the case until the advent of the first human, the first creature who knew that he knew. Thereafter continuing change becomes more and more by plan. The human being begins to take charge of evolution. There is still some chance and groping but much of it is really planned trial that results in either success or failure and we try again.

God, The Light, is the architect of all that is. His will, when manifested, becomes the purpose behind everything both the inanimate, the animate, and the thinking layers or spheres – of the universe.

In this e-Book's presentation of Teilhard de Chardin's Purpose Driven Evolution, the author uses chronological evolution as the outline as we observe the foundations of our universe and the unfolding of the phenomenon of "Man." The evolutionary process as explained in the Human Phenomenon, a new edition of de Chardin's *The Phenomenon of Man* (1965 English edition), as edited and translated from the French 1955 edition by Sarah Appleton Weber in 1999, is categorized under four parts:

I: Prelife
II: Life
III: Thought, and
IV: Superlife

Can a committed Christian believe in the concept (theory of Evolution)?

What does the Search for the Light in Evolution reveal about my place as a person in the development of the Earth? Read this e-Book to find the answers.

Purpose Statement

I have written this e-Book to help my readers answer the question, What does *Teilhard de Chardin: The Search For The Light In Evolution* reveal about my place as a human person in the development of the earth?

According to Teilhard, in order for the human person to see his/her real place in the development, he/she must be able to see "correctly."

The contents of the "Prologue: Seeing" topic through "The Second Basic Point of Teilhard's Process of Evolution," specifically detail what "correct seeing" is.

The Introduction to Teilhard's Purpose Driven Evolution, Part I Prelife, Part II Life, Part III Thought, and Part IV Superlife complete his description of the place of the human person in the development of earth.

Bernard J. Fleury Ed.D.

Preface

The first e-Book and Audio Book in the Called into Life Series is *How Jesus Christ Leads Us To The Kingdom Of Heaven*. We read or heard and pondered the many inspirational Bible verses in both the Old and New Testaments on life and light. We discovered how life and light in the Bible are really tied together (lifelight), with The Light, Jesus Christ, being the guide and the sustainer in life's journey of Spiritual Growth in Christian Living. Union with The Light in the heavenly Jerusalem, the Kingdom of Heaven is the ultimate fulfillment, the goal of every human life.

In the second e-Book and Audio Book in the Light Series, *The Mind-Body-Spirit Connection In The Medicine Of Light,* the cover image tells us the scope of this e-Book. The "head" represents the seat of the soul, the spirit, the inner light.

Nearly every form of the outer and inner light is to be found in this e-Book and Audio Book.

It is truly a self-help and self understanding guide to a healthier life.

In this third e-Book and Audio Book in the Light Series we ask the question, The What does *Teilhard de Chardin: The Search For The Light In Evolution* reveal about my place as a human person in the development of the earth.

We flesh out the following:

Man (male and female), the most highly developed consciousness, the leading shoot of evolution, the arrow pointing the way for further evolutionary development toward Omega Point: the final fulfillment of the world through creative union with Omega (God).

Teilhard's special words and phrases are each defined

Teilhard de Chardin The Search For The Light In Evolution

for the reader in the Simple Common Use Terms list on the next page.

Bernard J. Fleury Ed.D.

Teilhard's Special Words and Phrases in Simple Common Use Terms

1. Seeing: Seeing correctly is essential if we are to understand the universe as it is. In order to see correctly, we have to see ourselves, the human being, as the "center of perspective." Cf. Prologue: Seeing.

2. Cosmogenesis: The Universe in the process of becoming – its evolution.

3. Anthropogenesis: The evolution of the human phenomenon – the human person.

4. Natural units: e.g. A pile of sand is not a natural unit but a grain of sand is.

5. The "within" of things: every form of psychism – consciousness.

6. The "without" of things: their external structure – bones, brain, nervous system.

7. Psychic Energy: The basic, ultimate energy in the Universe.

8. Energy: A force that causes things to happen.

9. Noosphere: The sphere of thought that begins with the advent of man – the first creature who knows that he knows – reflective thoughts – able to observe and critique one's own behavior. The world of communication – the internet is a modern example of a world-wide electronic system that builds the noosphere – the sphere of thought.
Omega – supreme consciousness centered in one Divine Being who is three Devine Persons – A Trinity (Christianity).

In the beginning was The Fire, Alpha, God, and at the end is Omega, The Light, God.

Bernard J. Fleury Ed.D.

Pierre Teilhard de Chardin: The Man Who Christianized Purpose Driven Evolution

Who was this man who had the capacity to see all as part of one, to combine a basic scientific theory of evolution with a philosophical and theological vision? The son of a gentleman farmer and archivist, Pierre was born in 1881 in Auvergne, France. At the age of eighteen he decided to join the Jesuits by whom he had been educated for eight years. He spent his initial preparatory years in France and Jersey and later was sent to teach in Cairo, Egypt, for his three-year scholasticate. Returning to Sussex, England, he completed his four years of theology, and was ordained a priest in 1912. It was during his final preparatory years at Sussex that his profound interest in the theories of evolution began.

Following his ordination he returned to Paris for further preparation in geology. While in Paris, two events occurred which were to shape the rest of his life: (1) he met the Abbe Breuil who was to become his lifelong friend and colleague, and (2) his general interests in the theories of evolution were refined and centered on the evolution of man. From this point on, each of his life's experiences contributed something to the development of his theory of the evolution of man. He saw everything, even war, as part of the evolutionary process. His own experiences as a stretcher-bearer during World War I gave him firsthand knowledge of the horror of war; but nevertheless he saw in this cataclysmic upheaval a tearing down of the old and a building of the new – the conqueror assimilating but also being assimilated by the conquered.

After the "Great War," he spent several years in Paris earning his doctorate at the Sorbonne.

From 1923 to 1924 he went to China where he spent a year as part of a famous geological expedition that took him through much of Mongolia and northwestern China. His synthesis of philosophy, science, and religion revealed in his developing evolutionary theory of man was expressed in his *Letters from a Traveler* and *The Mass on the World.*

He returned to France to be greeted by the first of a series of official rejections of his teachings by his order and later by his church. An expression of his, made later in his life, sums up what his attitude was throughout these trials: "If what I have written is true it will remain. If it is not it were better forgotten." With these thoughts he continued to write though he could not publish, and what is an essential mark of his greatness, he continued to love and faithfully keep the commitments he had made to his religious order and his church.

Returning to China in 1926, he spent the next twenty years of his life there with brief visits to France, the United States, Abyssinia, Java, Burma, and India. These excursions, his work in China, and particularly his complete isolation there for six years during World War II brought his thought to maturity. He wrote a number of works during this period, chief of which was *The Phenomenon of Man,* which contains the fullest elaboration of his theory of the evolution of man. It is a light based theory

He returned to France in 1946 to face more rejection by his order and church, though he was a recognized scholar and had been offered prestigious academic positions and awarded a number of honors and memberships in French scientific circles. His health failed in 1947 when he suffered a serious heart attack. Following his recovery, he visited the United States and became associated with the Wenner-Gren Foundation that sponsored his two trips to South Africa.

Finally, in 1951, he came to New York to spend the rest of his life working under the auspices of the Wenner-Gren Foundation, playing an important role in framing anthropological policy, and helping organize international symposia in anthropology. He had always had a particular interest in that part of theology dealing with death, resurrection, and the Parousia or second coming of Christ. It seems wholly fitting then that the Christ who was the center of his life should call Teilhard to undergo the metamorphosis of death on Easter Sunday, April 10, 1955, after a day spent in worship and a quiet walk through the gardens of Central Park.

"If what I have written is true it will remain. If it is not it were better forgotten."

Before his death, Teilhard had been prevailed upon to leave his writings to his secretary so that they could be published after his death when his order's and the Church's prohibitions no longer applied. Within ten years after his death, the publication of his works had made him a household word among theologians and philosophers as he had been and continued to be among scientists. It was soon realized in Christian circles that Teilhard had effected something that was desperately needed; namely, a modern believable synthesis between religion, philosophy, and science. He had shown modern man that there is no necessary conflict between science and faith; indeed that each was crippled without the other; for to be a modern man one needed what religion gave: faith in the ahead.

Prologue: Seeing

Teilhard asserts that seeing correctly is essential if we are to understand the universe as it is. In order to see correctly, we have to see ourselves, the human being, as the "center of perspective."

As modern science had its beginnings during the Renaissance, early physicists and naturalists thought they could objectively observe phenomena apart from the "habits of thought" or "conventions" which shaped their respective disciplines. The fact is that we cannot really separate subjects (us with our ways of looking at things), from the objects (things) we are observing and are in turn affected by them. Teilhard maintains that because we are human, that is, have a highly developed consciousness with the power of reflection, we are looking at some aspect of *ourselves* in everything we see. (*The Human Phenomenon*, pages 3-4)

We human beings are the center. The intersection of the "landscape" (the human phenomenon) "we are passing through." By virtue of a highly developed "without" (primarily our brain and nervous system), we have a highly developed "within" (consciousness, power to reflect inner light). So because of both our "biological quality and properties of thought," the human being is, as alluded to previously, "the center of perspective" and also the "center of construction of the universe," that is, the evolution of the universe (cosmogenesis) is the milieu (surroundings, environment) of which the evolution of the human phenomenon (anthropogenesis) is an integral part. Thus it is that when we look at ourselves, the human beings, we grow in our knowledge of the universe. We evolve as long as we focus our eyes correctly.

Focusing our eyes correctly required the development over time of what Teilhard calls a "series of senses" among which are the following:

1. The sense of spatial immensity—the great and the small, the universe and the atomic particle;

2. The sense of depth—what appears to be the thin sheet of the past is really an endless series of events occurring over immeasurable temporal distances;

3. The sense of number—the staggering number of elements involved in the slightest change in the universe. When one uses a pressure cooker, the changes within are imperceptible. But if one doesn't monitor both the time and the temperature of what is being cooked, one soon finds out that an almost infinite number of changes have been going on when suddenly the applesauce within spews out of the opening in the cover, blowing off the gauge and depositing a blob of sauce on the ceiling;

4. The sense of movement—development hidden in the very slow is also illustrated by the example of the pressure cooker; and

5. The sense of proportion—realizing difference in physical scale that separates dimensions and movements of atom from nebula, the tiniest from the immense. (*The Human Phenomenon*, page 5)

According to Teilhard, once we can get beyond the illusions of "smallness, plurality, and immobility" we will have no trouble in seeing ourselves, the human beings as the "summit of an anthropogenesis, which itself crowns a cosmogenesis."

Arthur Zajonc, professor emeritus from Amherst College in Massachusetts, wrote that the Lights of Nature and Mind are entwined lights: Vision, seeing, requires more than a functioning physical organ. Without an inner light, without a formative visual imagination we cannot see what is actually there, we are blind. The outer light of nature (physical, the without of things), and the inner light of mind (the within of things) are really inseparable. (*Catching The Light,* page 5)

Humans of every time period and culture have to be constantly involved in forming and reforming the world they *see* based on their own level of imagination. What a human *sees* depends on his/her physical environment, the light without, the light of nature, and the light within, the within-the-mind process, the inner light.

In the Twenty First Century we *see* differently than the ancient Greeks or Romans did more than 2000 years ago. The concept and reality of space and space travel are utterly new. Our culture and language has changed due to our physical environment and the inner light which interprets it.

What we "see" is determined by our culture and language as studies of various cultures and languages have discovered. Each culture sees the world, its textures and colors in distinctive ways. Thus the two lights of nature and mind have interacted to present differing worlds to different ages.

In e-Book 4 of the Called Into Life by the Light Series of e-Books and Audio Books, *Out of Darkness Into the Light*, we will examine "light" in reported near death experiences. We will note that the specific ways in which near death experiencers describe their experience of light may differ even *if* all are actually seeing the same object because their experience, that is, "sight" of it, is different based on their culture and language.

However, it is also true that the mechanistic scientific view of the world begins in the Sixteenth Century and everything in it continued to advance in the Twenty First Century. The Harvard biologist, David Hubel, helped advance this view of the brain as a physical machine with thought, emotion, etcetera, as simply states or functions of a physical brain. We no longer needed illusions or mystical life forces. The brain is the sole reality. The human being is simply a very complex stimulus response mechanism. The brain is the computer that processes the stimuli and then produces a print out in terms of behavior. B. F. Skinner developed this Behaviorist Theory into both a therapy and an educational philosophy that produced Behavior Modification and Programmed Learning as the tools for shaping behavior. Hubel and Skinners' models rest on the basic assumption that ultimate reality, the basic stuff of the universe is physical in nature, matter of some sort, something that has mass (occupies space and has weight) and behavior. The Spiritual is simply one type of physical brain/nervous system behavior—nothing more. Everything about the human being is ultimately physical in nature.

We must be sure that what we conceive of as light really is light and not darkness! In Matthew's gospel we read: "The eye is the body's lamp. If your eyes are good, your body will be filled with light; if your eyes are bad, your body will be in darkness. And if your light is darkness, how deep will the darkness be?"

We must take responsibility for our models/perceptions. Do they foster the common good or are they selfishly "good" only for us? Our values matter because they influence the meaning of the content that we construct based on the world's presentation to our senses. But whatever kinds of images we fashion on our experience are just that, images only, a part of reality but not the whole of it.

Each image helps us to understand some aspect of reality; each domain of human experience plays its part in giving us a more complete (but never fully complete) understanding of reality. To gain the most complete picture of reality we can achieve is the result of the "inner light" (meaning) we achieve from a synthesis, of the images gained from all relevant disciplines, scientific, artistic, philosophical, and theological, focusing on a specific form of reality.

For example, let us look at a living tree. The artist sees it in terms of color, form, etc., sees it as a picture. The botanist classifies it by genus and species, examining the specifics of its trunk, branches, and leaves. The mathematician measures its diameter, circumference, and height. The philosopher sees the tree as a symbol—"the tree of life." All these disciplines contribute their special insight to our imaging of the reality we designate as a tree. We come to the fullest possible "light" about the tree and minimize our darkness (ignorance, false images) of it.

The prophet Isaiah cautions against mistaking light for darkness and false images:
> "Woe to those who call evil good, and good evil, who change darkness into light, and light into darkness" (Is. 5:20).

The importance of an inner light that is truly light and not darkness is a repeated scriptural theme. God the Father creates light out of chaos and darkness. In a hymn written by John Marriott we note the role of Jesus and the Holy Spirit.

> Savior, who came to bring
> On your redeeming wing
> Healing and sight;
> Health to the sick in mind,
> Sight to the inly blind;
> O now to all mankind
> Let there be light.
> Spirit of truth and love,
> Life-giving holy dove,
> Speed on your flight;
> Move on the water's face,
> Bearing the lamp of grace,
> And in earth's darkest place
> Let there be light.

It is vital for our own individual lives and for the world we live in that we continue to struggle to discern light from darkness.

Bernard J. Fleury Ed.D.

Teilhard de Chardin and Arthur Zajonc Synthesizers

In my opinion, De Chardin would certainly agree with Zajonc that the lights of nature and mind are inextricably entwined. Arthur Zajonc's book, *Catching the Light* is a fascinating treatise on the very concept of Light. Although he is a professor of physics and a specialist in quantum physics, his work goes much beyond what is conventionally thought of as the realm of physics. His approach to Light is much like Teilhard de Chardin's approach to ultimate reality – he is a synthesizer who has the gift of being able to bring to bear the insights of history, science, mathematics, religion, and art on the Concept of Light from biblical times to the modern era.

I first discovered Teilhard de Chardin's view of directed evolution when a teacher at the Elementary School where I was the Principal, gave me a copy of his book, *The Phenomenon of Man*, Second Harper Torchbook Edition, published in 1965. I read it several times, cover to cover. Thus began my introduction to his views on evolution and humankind. I have read and taught as a College Professor from many of his major works.

When I discovered Arthur Zajonc's book *Catching the Light*, published by Bantam Books in 1993, my interest in "Light" had grown and I saw in Zajonc's book a mirror of all that Teilhard also believed and more.

In this e-Book *Teilhard de Chardin Searching For The Light In Evolution,* I focused on the role of light in evolution from Teilhard's point of view.

I used Wikipedia as the primary source with confirmation of the facts I used from a highly respected Author, Arthur Zajonc in his book *Catching the Light.*

I highly recommend that my readers purchase Zajonc's

book. They are in for a treat. They will read it cover to cover and likely more than once.

The outer light is incomprehensible to the human eye without various levels of development in the inner light of the mind. All of the various concepts, explanations, of "light" are the direct result of the interaction of the inner light of the mind through the various sense organs of the body. The external light (stimuli) affect the development of the inner light (mind) and the inner light shapes the ways in which external light is received and perceived. In his commentary on the evolution of matter in Chapter One, "Before Life Came," in *The Phenomenon of Man*, Teilhard de Chardin writes: "To begin with at the very bottom there is a still unresolved simplicity, luminous in nature and not to be defined in terms of figures." (*Phenomenon of Man*, page 47)

One of the reasons Sarah Appleton-Weber gives for renaming *The Phenomenon of Man* as *The Human Phenomenon* instead, is because the concept of Human Phenomenon is much broader and more in keeping with Teilhard's view. For we humans to see ourselves completely and correctly, we have to see our individual selves as part of humanity, humanity as part of life, and life as part of the universe. To emphasize this, Teilhard organizes his work under four headings: pre-life, life, thought (the past), and super-life (the future). (*The Human Phenomenon*, page 5)

The human being is a fact of nature, a true phenomenon, not an epiphenomenon or freak occurrence. This can be lost if we overstress human individuality (the phenomenon of man), at the expense of the human as a collective (the human phenomenon). The human being is a composite of the "within" and the "without" as is the entire universe. Thus it is that the human is the axis, the leading shoot, the arrow pointing the way of evolution in the direction of ever higher forms of consciousness culminating in a final union: Omega Point. (*The Human Phenomenon*, page 6)

Teilhard maintains that though it's true that the phenomenon of consciousness has been restricted by most scientists to higher forms of life and is most evident in humans, we cannot assume that this phenomenon is not present to some degree in other forms of life. In fact, it is Teilhard's insistence on the inclusion and emphasis on the "within" (consciousness) as the core of the phenomenon of man, and indeed of the entire universe, that gives Teilhard's thought its distinctive character.

The First Basic Point of Teilhard's Process of Evolution

According to Teilhard the basic substance of the universe, luminous in nature, is energy and essentially all energy is psychic. While leading the reader to this answer he presents two other basic points of his thought:

1. There is a "within" (every form of psychism-consciousness) and a "without" of things (external structure – brain, bones, nervous system) – this quality of the earth was present from the very beginning.

The "within" and the "without" of the human species continues to evolve. As we grow and change, so too do our priorities and interests from which our searching questions regarding nature, flow. It is a fact that the value of the answers we get depends on the kinds of questions we ask. Nature gives most of her evidence in answer to the questions we ask her. So, the nature of the questions we ask largely determines the kinds/nature of information we get in answer.

If we want to get as complete a picture as possible of the human attempt to comprehend light, we need to be open to all the insights of the past as well as to our modern day notions. The search for a final, comprehensive, understanding of the nature of light goes on and we're nowhere near exhausting the concept (s) of light and probably never will be. If light is Someone and not merely something—and if that Someone is the Divine, Limitless, Infinite source of all that is, we never will be able to completely understand light as it is, because our finite minds can never fully fathom the infinite mind of God, the Light.

The existence of two separate, equal, uncreated, creating principles, good and evil, light and darkness, (Manichaeism) was anathema to the Apostolic Catholic Church. God is light, the *sole* uncreated creator. Darkness was viewed as a negation, a lacking, the absence of light. Darkness and evil entered the world through the rebellion of one of the most brilliant of the angels, Lucifer, the light bearer, who refused to obey God's commands and was thrust out of heaven and became the Prince of Darkness. He became and is the source of darkness in human lives, anxious to have others join him in his misery.

But light is the Supreme Good, not darkness. Lucifer, the evil one, is a creature, God, the Light, is the sole creator. Light will always triumph in the end—so the Fall of Man necessitates a redeemer who will restore light and life. For Christians, Jesus Christ, Light from Light is that redeemer.

But the Persian Philosopher Mani's ideas survived for centuries. The Cathars or "Pure Ones" were a French offshoot of Mani. They shared his view that the physical world was the creation of Satan. Through rigorous fasting and penances they believed that the divine light within them, their spirits, could be saved from the corruption of the body. But if they didn't make it the first time, reincarnation for a second chance was possible.

The Cathars were rigorously condemned because they preached the evil of the body and the Catholic Church believed and taught that all of God's creation, physical and spiritual was good. In fact, "holiness" really meant true "wholeness" that is establishing and maintaining a right relationship between the two *good* aspects of human existence, the body and the spirit.

At about the time that the Cathars were being wiped out, Robert Grosseteste, Bishop of Lincoln in England, and by invitation, instructor in the formation of Franciscan seminarians at a new school, Oxford, was producing his book *De Luce, (On Light)*. Grosseteste's story of the origin of the universe is remarkably like that of Teilhard de Chardin who wrote his story more than seven centuries later. In the beginning was the Light, God, complete in Himself. At the point of Creation His light expands and assumes two forms, the corporeal (physical, "without"), and the incorporeal (spirit, "within"). Grosseteste characterizes this spiritual light as personal because it is manifested in intelligent, purely spiritual beings, the angels, and is part of man himself.

For both Grosseteste and Teilhard, metaphysics and the physics of light, especially mathematics for Grosseteste, were interrelated and essential components of God's creation. Grosseteste saw God as the Great Mathematician, who forms (Geometer) and gives matter its characteristics of number, weight, and measure (Numerator).

Reading Grosseteste is like reading the works of Teilhard de Chardin. Both believed in the centrality of light and the importance of both the physical and the spiritual. No one human study standing alone can give us any sort of comprehensive picture of reality—each study (history, philosophy, math, science, theology) sheds "light" on part of the picture. We need their joint insights to really see.

The Body (Mass) Called Light

Aristotle saw light as a condition or a state of a medium – not a "thing" with a structure.

Isaac Newton, January 4, 1643 – March 31, 1727, received his Master of Arts from Trinity College in 1667 and became a Fellow there. By 1669 he was Lucassian Professor of Mathematics. He became fascinated with "light" and "colours".

During the 1670-1672 period he lectured on optics. He investigated the refraction of light, demonstrating that a single ray of white light could be decomposed through a prism into the colours of the spectrum, the rainbow. White light (the color "white") is a combination of all the separate colors of the rainbow, the spectrum.

In 1704 Newton published Optics in which he stated his corpuscular theory of light. He considered light to be made up of extremely subtle (very small) corpuscles (bodies) that ordinary matter was made of grosser (larger) corpuscles. He asked "Are not gross (large) Bodies and Light convertible into one another....and may not Bodies receive much of their Activity from the Particles of Light which enter their Composition?" (Wikipedia Contributors, "Isaac Newton", Accessed July 17, 2015).

Arthur Zajonc confirms in *Catching The Light,* Chapter 5, that Light was Matter, a physical thing. The cosmos, as Teilhard would later write, was one unified phenomenon, from the immense to the most-minute. All moved by laws that Newton had discovered.

Was Rene` Descartes Dream a Vision or an Illusion?

Rene` Descartes, March 31, 1596 to February 11, 1650. Descartes was known to me in Philosophy for his famous saying "Cogito ergo sum". I think therefore I am. The act of thinking requires a thinker. His ideas on light are similar to his ideas on thinking. Light is a mechanism like thinking. In the case of light, it is a physical mechanism, an action between my eye and the object I am looking at.

Newton would build on Descartes concept and change it! (Wikipedia Contributors, Rene` Descartes, accessed July 17, 2015).

Leonard Euler, April 16, 1707 to September 18, 1783. Euler disagreed with Newton's corpuscular theory of light in the Opticks, which was then the prevailing theory. His 1740s papers on optics helped insure that the wave theory of light proposed by Christian Huygens would become the dominant mode of thought, at least until the development of the quantum theory of light. Euler thought that luminous (radiating or reflecting light) objects vibrate (moved to and fro, as a pendulum, or up and down, quickly and repeatedly) and that the "ether" carries these vibrations to the eye as Euler said air carries sound to the ear. (Wikipedia Contributors, Leonard Euler, accessed July 17, 2015). Zajonc confirms these facts in *Catching The Light,* Chapter 5.)

Robert Boyle, January 25, 1627 to December 31, 1691. Boyle was the youngest of fourteen children born to Richard Boyle and Catherine Fenton. Richard was an enormously wealthy landowner in Ireland. Robert became a scientific investigator of some merit. Among the important work he accomplished in physics was the discovery of the part taken by air in the propagation of sound using the air pump he invented to create a vacuum. As a result of his experiments with the propagation of sound in a vacuum (no air) had concluded that air was necessary for the transmission of sound. Further experiments with the transmission of light led him to conclude that air was not necessary for the transmission of light. (Wikipedia Contributors, Robert Boyle, accessed July 17, 2015. Confirmed by Zajonc in *Catching The Light,* Chapter 5.)

Wave Theory of Light

Thomas Young, June 13, 1773 to May 10, 1829.

In Young's own judgment, of his many achievements the most important was to establish the wave theory of light. To do so, he had to overcome the century-old view, expressed in the venerable Isaac Newton's "Optics", that light is a particle. Nevertheless, in the early 19th century Young put forth a number of theoretical reasons supporting the wave theory of light, and he developed two enduring demonstrations to support this viewpoint. With the ripple tank he demonstrated the idea of interference in the context of water waves. With the Young's interference experiment, or double slit interference, he demonstrated interference in the context of light as a wave. He also subscribed to the Concept of a luminiferous ether "which was subsequently rejected by his critics.

"The experiments I am about to relate ... may be repeated with great ease, whenever the sun shines, and without any other apparatus than is at hand to everyone."

This is how Thomas Young speaking on 24 November 1803, to the Royal Society of London, began his description of the historic experiment. His talk was published in the following year's Philosophical Transactions, and was destined to become a classic, still reprinted and read today.

In the subsequent paper entitled *Experiments and Calculations Relative to Physical Optics*, published in 1804, Young describes an experiment in which he placed a narrow card (approx. 1/30th in.) in a **beam of light** from a single opening in a window and observed the fringes of color in the shadow and to the sides of the card. He observed that placing another card before or after the narrow strip so as to prevent light from the beam from striking one of its edges caused the fringes to disappear. This supported the contention that light is composed of **waves**. Young performed and analysed a number of experiments, including interference of light from reflection off nearby pairs of micrometre grooves, from reflection off thin films of soap and oil, and from **Newton's rings**. He also performed two important diffraction experiments using fibres and long narrow strips. In his *Course of Lectures on Natural Philosophy and the Mechanical Arts* (1807) he gives **Grimaldi** credit for first observing the fringes in the shadow of an object placed in a beam of light. Within ten years, much of Young's work was reproduced and then extended by **Fresnel**. (Tony Rothman in *Everything's Relative and Other Fables from Science and Technology* argues that there is no clear evidence that Young actually did the two-slit experiment. *See also Newton wave-particle duality.*) (Wikipedia Contributors, Thomas Young (Scientist), accessed July 17, 2015. Confirmed by Zajonc in *Catching The Light,* Chapter 5,)

Augustin-Jean Fresnel, May 10, 1788 to July 14, 1827

Augustin-Jean Fresnel was a French engineer and physicist who contributed significantly to the establishment of the theory of wave optics. Fresnel studied the behavior of light both theoretically and experimentally.

In 1818 he published his Memoir on Diffraction of Light, submitted it to the Academe of science in the same year. (Diffraction, modulation of waves in response to an object or slit, in the path of propagation, giving rise in light waves to a banded pattern or a spectrum, like a rainbow.) His discoveries and mathematical deductions, building on experimental work by Thomas Young, extended the wave theory of light to a large class of optical phenomena. (Wikipedia Contributors, Augustin-Jean Fresnel, accessed July 17, 2015. Confirmed by Zajonc in *Catching The Light*, Chapter 5.)

Francois Arago, February 26, 1786 to October 2, 1853

Dominique Francois Jean Arago, known simply as Francois Arago was a French mathematician, physicist, astronomer, freemason, supporter of the carbonari and politician. Arago warmly supported Augustin-Jean Fresnel's optical theories, helping to confirm Fresnel's wave theory of light by observing what is now known as the spot of Arago. The two philosophers conducted together those experiments on the polarization of light which led to the inference that the vibrations o the luminiferous ether were transverse to direction of motion, and that polarization consisted of a resolution of rectilinear propagation into components at right angles to each other. The subsequent invention of the polariscope and discovery of Rotary polarization and are due to Arago. He invented the first polarization filter in 1812.

In optics, Arago not only made important optical discoveries on his own, but is credited with stimulating the genius of Jean-Augustin Fresnel, with whose history, as well as that of Etienne-Louis Malus and Thomas Young, this part of his life is closely interwoven.

Shortly after the beginning of the 19th century the labours of at least three philosophers were shaping the doctrine of the undulatory, or wave, theory of light. Fresnel's arguments in favour of that theory found little favour with Laplace, Poisson and Biot, the champions of the emission theory: but they were ardently espoused by Humboldt and by Arago, who had been appointed by the Academy to report on the paper. This was the foundation of an intimate friendship between Arago and Fresnel, and of a determination to carry on together further fundamental laws of the polarization of light known by their means. As a result of this work, Arago constructed a polarscope, which he used for some interesting observations on the polarization of the light of the sky. (Wikipedia Contributors, Francois Arago, accessed July 17, 2015. Confirmed by Zajonc in *Catching The Light*, Chapter 5.)

Newton, Desecartes, Young, and Euler had at least one feature in common. They all viewed light as something "other", matter, waves, vibrations, etc. Was light a wave or a vibration? Maxwell and Faraday answer that light was a wave, but what is this wave composed of? What kind of wave is it?

The Light of Electricity: A Way of Seeing

In 1791, **Michael Faraday** was born to a poor London blacksmith. He had the barest rudiments of schooling in the three "R's", but by the time of his death in 1867 he had become one of the greatest, if not the greatest experimental scientists of all time. He was a deeply Christian man who believed that his scientific experiments with the visible world led to a better understanding of things unseen. Through his contemplation of nature he saw nature's God, so his theology and philosophical science were two sides of a single reality for him. He was much like Teilhard de Chardin in the way he viewed the universe.

At the age of thirteen Faraday became an apprentice bookbinder, a trade that gave him ready access to books and learning he would not have had otherwise. Upon completing his apprenticeship at the age of twenty-one he secured a position as assistant to Sir Humphrey Davey of the Royal Institution, the most famous scientist in England of his day. Faraday actually became assistant to all the scientists at the Institution, and then began to strike out on his own with experiments that led him away from the materialistic constraints of his contemporaries.

Because he believed in the basic unity of nature with its creator, he subscribed to the view later promulgated by Teilhard, that what appeared on the surface to be an isolated instance, an epiphenomenon, was really spread throughout the world—a phenomenon that just happened to clearly manifest itself in a particular experiment. (Facts about Faraday, his life and experiments taken from Wikipedia Contributors, Michael Faraday, accessed July 17, 2015. Confirmed by Zajonc in *Catching The Light*, Chapter 5.)

The similarities between Michael Faraday's and Teilhard de Chardin's approach to Science as Natural Philosophy, the conviction that Science and God, true faith and valid reason, *do not* conflict, that the study of the visible leads to the invisible (God) and that nature is a unity, is remarkable. One of the stories about the events that led Teilhard de Chardin to God and his vocation as a priest-scientist relates that even as a young boy Teilhard was always searching for something that was permanent. One autumn he stumbled across an old iron plow in a field. He kicked it—it was solid.

Ah, at last he had found something that was permanent—something that would last! But the next spring when he saw the plow again, it was covered with rust—sign of decay. Thus began his search for the permanent, the lasting, the ultimate, through his investigations of the visible, impermanent, material world.

As we just noted, Teilhard also believed that because of the unity of the natural world we often find the universal hidden beneath the exceptional. (Pierre Teilhard de Chardin, *The Phenomenon of Man*. page 56.)

The example of radium is a good one. Radioactivity was observed in radium—what would have happened to modern physics if this was considered to be a function of an abnormal substance rather than the key which unlocked the secrets of nuclear physics— a common property of the entire universe—but at first perceptible only in radium?

Faraday's conception of electromagnetic induction was as transformative of our scientific image of light as was the discovery of radioactivity through the observations of the behavior of radium.

As Faraday believed, the character of light seemed to be wavelike. As a result of his experiments he came to the conclusion that a "wave of electricity" is caused by sudden changes in the current through a first or "primary" circuit. This electrical wave travels through space and induces a similar disturbance in the nearby "secondary" coil of wire. (Which was registered by the movement of a needle on an attached Galvanometer)...yet just what is this "electric wave" that it can connect distant circuits without a visible material connection of any kind? Remember, the concept of a material-ether that served as a pathway, was dead. His answer resulted in a new scientific image of light. (Wikipedia Contributors, Michael Faraday, Accessed July 17, 2015. Confirmed by Zajonc in *Catching The Light*, Chapter 6.)

Faraday believed that direct experience was the best source of truths so he distrusted theories that were purely speculative.

So we "know" things by their attributes, that is, their observable characteristics. Faraday suggested, "force, not substance is the true being of the world." Forces may group themselves in many different ways to form our observable material world, but at the heart of things is "force," energy, light not matter (substance). Wikipedia Contributors, Michael Faraday, accessed July 17, 2015. Confirmed by Zajonc in *Catching The Light*. Chapter 6.)

Once again, Faraday's views of "Force" as the true being of the world and Teilhard de Chardin's view that energy (a force which causes things to happen) are strikingly similar. Teilhard posits two energies at work in the Universe: tangential energy (the "without" or physical aspect of things) and radial energy (the "within" or psychic aspect of things). These two energies are forces that cause things to happen. Ultimately Psychic Energy is at the heart of the universe— the beginning and core of all creation. To Teilhard, Psychic Energy centered in a Person is God, the foundational force of all that was, is, or ever will be.

Light and Magnetism

James Clerk Maxwell, June 13, 1831 to November 5, 1879

James Clerk Maxwell, son of a distinguished family in Scotland and a mathematician, took up the results of Faraday's experiments and thinking.

Maxwell is credited with being a discoverer of the laws of electrodynamics. Like Teilhard de Chardin. Maxwell was a synthesizer. He brought together the separate "knowledges" of electricity and magnetism and expressed this now unified knowledge in his four "Maxwell" equations.

Maxwell's investigations unified the studies of light and electromagnetism. His paper On Physical Lines of Force – written over the course of two years (1861-1862) and ultimately published in several parts – introduced his pivotal theory of electromagnetism. Among the tenets of his theory were (1) that electromagnetic waves travel at the speed of light, and (2) that light exists in the same medium as electric and magnetic phenomena. The most crucial findings of his electromagnetic theory – that light is an electromagnetic wave, that electric and magnetic fields travel in the form of waves at the speed of light, that *radio waves* can travel through space – constitute his most important legacy. Nothing sums up the monumental achievement of Maxwell's life work as well as these words from Einstein himself: "This change in the conception of reality is the most profound and the most fruitful that physics has experienced since the time of Newton."

Maxwell believed that these two natural phenomena were intimately related, as did Faraday. Subsequently he would conclude that both light and magnetism are affections of "force." Forces, not the speculative ether, were the true being of the world, but it would take decades before this view became the predominant one. Maxwell saw every- body as surrounded in space, not by a material-ether, but rather by an electromagnetic field in which the energy (force) resides. (Wikipedia Contributors, James Clerk Maxwell, accessed July 17, 2015. Confirmed by Zajonc in *Catching The Light,* Chapter 6.)

It is interesting to note that this entire theory of electromagnetic fields would subsequently play an important part in the descriptions of their out-of-body perceptions reported by persons who had undergone near-death experiences. There was an apparent "force" of the nature of light that drew them in a certain direction. It was a force that surrounded, as a halo, the dead that persons met in the near-death experiences. We will consider this phenomenon further in e-Book Four.

Bernard J. Fleury Ed.D.

The Second Basic Point of Teilhard's Process of Evolution

2. The law of "complexification": the stuff of the universe tends toward increased complexity as evolution progresses, and there is a direct and positive relationship between the complexity of the "without" of things and how well developed their "within" or consciousness is.

In the evolution of man, evolution went straight to work on the brain and nervous system. Humans have the largest brain and the best developed nervous system of any of the primates and therefore the highest and best developed consciousness, capable because of the light within, to see, that is, understand in greater depth and in varying ways, the light without.

This same theme of "Entwined Lights" can be found in the Judaeo-Christian Scriptures and Creeds.

In the Nicene Creed, Jesus is affirmed to be "eternally begotten of the Father, God from God, Light from Light, true God from true God." The evangelist, John, opens his gospel with:

In the beginning was the Word; the Word was in God's presence, and the Word was God. He was present to God in the beginning. Through him all things came into being and apart from him nothing came to be. Whatever came to be in him, found life, life for the light of men. The light shines on in darkness, a darkness that did not overcome it...The real light which gives light to every man was coming into the world. (John Chapter 1, Verses 1- 5 and 9).

Light is the basic Existent of the entire universe—not a something alone, but also a Someone – a something in nature, a Someone *who forms* and *informs* the human mind. The individual human mind has to be developmentally ready to both receive and perceive the light for the light to be comprehensible. This is a true state of affairs for both the light of nature and the light of mind. Human freedom and choice in terms of willingness to receive the light (the teachings and recognition of who Jesus is) is the essential prerequisite to Faith. Faith in turn is a prerequisite for receiving the gifts of the Holy Spirit, among which are Knowledge, Wisdom, and Understanding—all gifts of enlightenment.

Introduction to Teilhard's Purpose Driven Evolution

In his Introduction to the 1959 edition of *The Phenomenon of Man*, Sir Julian Huxley asserts that Teilhard's influence on scientific and theological thinking has to be important because Teilhard was a great synthesizer in the mold of Thomas Acquinas. Science and religion, the material and spiritual worlds, had to be seen as two aspects of the evolution of the entire universe including humankind. The attempt to keep science and religion in completely separate, thought-tight compartments only serves to continue to promote a false vision of reality, a distorted vision of the universe.

In her 1999 translation of Teilhard's work, Sarah Appleton-Weber maintains that even the title *The Phenomenon of Man* is incorrect; so she renames her edited version *The Human Phenomenon*. The 1959 error, according to Appleton-Weber was due to the failure of the translators to note the differences between two French words "homme" and "humain." "Homme" refers only to man in his physical aspect, his material body. In this dimension "man," the "human being" is insignificant, only one physical being among many. But "human" refers both to the "within" and the "without" the psychic (spiritual) and the physical (body). Seen as a coherent (integrated) whole, Teilhard asserts that the human is indeed the axis and the arrow pointing the way to the direction of evolution itself.

Our very survival depends on each of us seeing that phenomena are never stationary but always processes or part of processes. In fact, as a whole a phenomenon is always in the process of becoming, of attaining new and more complex levels of existence and organization which Teilhard called a genesis or evolution.

But the process is a *Purpose Driven* evolution. In the beginning was The Fire—The Big Bang—the physical world thrust into existence and motion by The Light. All of creation is *Purpose Driven*.

In the Beginning Was The Fire

God, The Light, the source of all other forms of light in the Universe both physical the light without, matter, the body, and psychic, spiritual, the inner light. The "within" and "without" of all natural units (the human being), are the entwined lights of nature and mind.

"The earth was a formless, wasteland and darkness covered the abyss, (Genesis Chapter 1, verse 2).

The cosmic explosion, the big bang, a huge fire occurs. Hot matter and radiation are created which expand and cool to form the galaxies and stars observed today.

But who or what started the big bang? Was it how time and space began?

Darwin's thesis that evolution proceeds by chance, groping, and the survival of the fittest could be the case until the advent of the first human, the first creature who knew that he knew. Thereafter continuing change becomes more and more by plan. The human being begins to take charge of evolution. There is still some chance and groping but much of it is really planned trial that results in either success or failure and we try again.

God, The Light, is the architect of all that is. His will when manifested becomes the purpose behind everything both the inanimate, the animate, and the thinking layers or spheres of the universe.

Light is our origin, our mainstay, and our destiny. It is what forms and informs us. We are called into life by The Light, we live in the light, and our destiny is The Light.

Light: Foundational Energy of the Universe

In this presentation of Teilhard's purpose driven evolution, we shall use chronological evolution as the outline as we observe the unfolding of the phenomenon, "Man." The evolutionary process as the phenomenon, "Man." The evolutionary process as explained in *The Human Phenomenon* (HP), a new edition of *The Phenomenon of Man* (PM) [1955 in French, 1959 and 1965 in English editions, as edited and translated by Sarah Appleton- Weber in 1999] is categorized under four basic parts:

I: Prelife
II: Life
III: Thought, and
IV: Superlife

We will examine Parts I to IV to discover Teilhard's answers to the questions "What is Man?" and "What is the foundational energy of the universe?" Comments within parentheses indicate my perceived connections of Teilhard's thought with light and "The Light" (God).

Part I: Prelife

What is the basic stuff of the Universe? What is the ultimate or real? Teilhard begins his answer to these questions by writing, "Elementary Matter...the stuff of tangible things" that "reveals itself with increasing insistence as radically particulate, yet basically connected, and finally, prodigiously active. Plurality, unity, energy: The three aspects of matter" (*The Human Phenomenon*, page 12).

Bernard J. Fleury Ed.D.

Let us look at these three aspects of matter in more detail— perhaps one aspect is the mirror in which the other two are simply reflections.

Elementary Matter reveals itself to be:

1. radically particulate (plurality)—bewildering in its multiplicity and its minuteness, the substratum of the tangible universe is in an unending state of disintegration, endlessly degrading into smaller and smaller particles as it goes downward (*The Phenomenon of Man*, page 41 and *The Human Phenomenon*, page 12).

2. essentially related, that is, basically connected (unity)— the more we split and pulverize matter artificially, the more insistently it proclaims its fundamental unity....Molecules, atoms, electrons—whatever the name, whatever the scale— these minute units (at any rate when viewed from our distance) manifest a perfect identity of mass and of behavior. (*The Phenomenon of Man* page 41 and *The Human Phenomenon* pages 12-13).

3. prodigiously active (energy)—a capacity for action, or more exactly, for interaction. Energy is the measure of that which passes from one atom to another in the course of their transformations. A unifying power, then, a power of bond-ing but also, because the atom appears to become enriched or exhausted in the course of the change, the expression of structure (or building up)...Though never found in a state of purity, but always more or less granulated (even in light) energy nowadays represents for science the most primitive form of universal stuff. (*Phenomenon of Man*, page 42 and *The Human Phenomenon*, pages 13-14).

Energy is then the heart of matter, the basic component of

all that is. From the explanation given to this point, it would appear that the basic stuff of the universe is energy, energy that is physical in nature because it is subject to empirical verification and measurement. But for Teilhard, physical energy is not the ultimate, the basic reality, the *is*. We will have to follow along with him as he delves deeper into matter and its evolution to find his ultimate reality.

According to Teilhard, we cannot really understand any part of the matter of the universe apart from the total universe.

Considered in its physical, concrete reality, the stuff of the universe cannot be split apart but, as a kind of gigantic atom it forms in its totality (apart from thought on which it is centered and concentrated at the other end) the only real indivisible. The history of consciousness and its place in the world remain incomprehensible to anyone who has not seen first of all that the cosmos in which man finds himself caught up constitutes, by reason of the unimpeachable wholeness of its whole, a system, a totum, and a quantum: a system by its plurality (that is a grouping of a multiplicity of minute particles), "a totum by its unity" (the minute particles or units which make up the universe manifest a perfect identity of mass and of behavior—they're all alike), and "a quantum by its energy" (a certain amount of energy entrapped in universal space which according to Teilhard physicists already think they are in a position to measure) (*The Phenomenon of Man*, pages 43-46 and *The Human Phenomenon*, page 14).

This energy (quantum) takes on its full signification however, "when we try to define it with regard to a concrete natural movement— that is to say, in duration (*The Phenomenon of Man*, page 46) and we are thus led to consider the evolution of matter which follows a process that Teilhard labels complexification."

At the heart of the physical world is matter, something that occupies space and has weight. But before the simple corpuscles that formed original matter, there was what

Teilhard terms "a kind of luminosity" (light) "indefinable in terms of figures."

The simple original corpuscles, consisting of protons, neutrons, electrons, photons, group together in various ways to form the simple elements like hydrogen. Complexification continues and compound bodies appear which are multicellular. Finally, when these multicellular molecular masses reach a certain stage of complexification, a critical stage, life appears (*The Phenomenon of Man*, pages 47-48).

But, when we examine this process, if we only look at the exterior, the physical structure, the "without" of things, we only observe half of any phenomenon. If we want to study the entire phenomenon, we have to take a look at the "within," the "interior," the "reserve" of things. This is especially important as evolution moves toward the appearance of the human (*The Phenomenon of Man*, page 54).

Teilhard makes three assertions with regard to the "within" of things that are essential to his thought: that there is a "within," that this "within" has some connections/relationships with the "energy" recognized by Science that are either qualitative or quantitative. He presents his reasoning for these assertions in Chapter II, "The Inside of Things," under three subchapters: Existence, The Qualitative Laws of Growth, and Spiritual Energy.

We are still looking for his ultimate, his real—to find it we must continue along with him as he discusses "Existence."

Light enters the scene at the outset of Teilhard's treatment of "Existence." With the discovery of radioactive substances, the concept that the chemical elements we learned about in high school were stable over a long period of time, suddenly became an illusion. Nothing is really "solid"; everything is in constant flux when viewed over a long period of time (duration). The velocity of atomic movements, their *speed* drastically modifies a body's mass. It is only when movements are sufficiently slowed down that they *appear* to be solid or immobile. All "speed" is referenced to the speed of light, so every body is not only changed by its speed, but it also *radiates,* that is, it moves like rays…from a center, it emits rays, as of light or heat (*The Phenomenon of Man*, pages 54-55 and *The Human Phenomenon*, page 23).

What is true of the physical "without" of things is also true of the interior "within" of things.

Teilhard relates that up until the time of his writing (nearly mid- twentieth century), all that really existed for physicists was matter, the physical, the "without" of things. But as we move up the scale of complexification, and especially when we reach the realm of plants (botany), it becomes more and more difficult to maintain that the "without" of things is all there is. For that portion of biology which deals with the behavior of insects and coelenterates, the attempt to legitimately maintain that nothing exists but the "without" is nearly impossible. Finally, when we reach the advent of the human, we can't deny the existence of a "within" (an interior, a self-consciousness), because its existence is the result of a direct awareness or intuition, the most rudimentary form of knowledge from which all knowledge, all learning, all science, directly springs (*The Phenomenon of Man*, pages 54 -55 and *The Human Phenomenon page* 23).

Teilhard continues his line of thought regarding the existence of the "within of things", the inner light, consciousness, with the assertion that far from being an epiphenomenon, a queer exception to be found only in man, consciousness is rather a phenomenon, indeed the central and most essential phenomenon of the universe. Just because it appears fully developed only in man does not mean that consciousness is restricted to man alone— that it is not present elsewhere in the universe. For the exceptional (what we've often termed as an "epiphenomenon"), actually reveals something that is universal. This has happened far too often for us not to admit it.

Because of the "fundamental unity of the world," (light is at the heart of and part of all things), what we correctly observe (even if only in one spot), is in some way or other present everywhere. The example of radium is a good one. Radioactivity was observed in radium. What would have happened to modern physics if this was considered to be a function of an abnormal substance rather than the key which unlocked the secrets of nuclear physics, a common property of the entire universe, but at first only perceptible in radium (*The Phenomenon of Man*, page 56)?

When we apply this concept of the universal hidden beneath the exceptional, where does our human self-knowledge lead us? It leads us to realize that though consciousness, especially self-awareness, is completely evident only in the human, this does not mean that it is limited to human beings. In fact, just the opposite is indicated. For just as in the case of radium, an isolated phenomenon really indicated a reality that is diffused throughout the universe because of the fundamental unity of the world.

The existence of a "within" to things from the very beginning of time, with the term "within" taken in its widest sense, that is, every kind of psychism from the most rudimentary to human reflective thought, is the case. Even primitive matter, a mechanical layer, has hidden within it the potential for an "interior," a "within," which is, due to the simplicity of the "without," the physical structure, undeveloped, but still there.

So, what is required for this "interior" or "within" to develop, to become completely perceptible?

According to Teilhard, the answer is that it develops according to certain rules that he calls "The Qualitative Laws of Growth" which he cites under three "observations."

In the first observation Teilhard refers to what I have just cited regarding "primitive" or as Appleton-Weber calls it "nascent" matter. The farther back we go, this matter reveals itself to be composed of innumerable tiny particles which are seemingly all alike, spread throughout (coextensive with) the entire universe and somehow connected to each other by a "global" or "energy of the whole" (*The Phenomenon of Man*, page 59 and *The Human Phenomenon*, page 26).

The second observation indicates that the elements of consciousness became more evident as complexification continues in the process of evolution. As complexification progresses, the elements of primitive matter tend to differentiate their nature, that is, take on specific characteristics.

In his third observation, Teilhard repeats this theme that the human phenomenon has two interconnected aspects, a "without" and a "within." Complexification of the "without," the physical structure, always has an effect on the development of the "within" or consciousness. (*The Phenomenon of Man*, pages 59-60 and *The Human Phenomenon*, pages 26- 27).

But how do we hold together spirit (the "within") and matter (the "without")? What kind of research and interpretation should an integral science of nature propose if we are to show that matter and spirit are both interdependent and independent?

Teilhard's answer is that the basic energy of the universe is psychic but that this energy takes on two forms: a "tangential energy" that links all elements of the same order (complexity and centricity), and a "radial energy" which pushes the complexification and centricity processes to ever greater complexity and centricity. (*The Phenomenon of Man*, pages 63-65 and *The Human Phenomenon*, page 29).

Finally, Teilhard has led us to his answer to the question "What is ultimate or real?" when he says that the basic stuff of the universe is energy and that essentially all energy is psychic. While leading us to this answer, he has also presented two other basic points of his thought:

1. There is a "within" and a "without" of things. This quality of the earth was present from the very beginning.

2. The law of "complexification":

The stuff of the universe tends toward increased complexity as evolution progresses and there is a direct and positive relationship between the complexity of the "without" of things and how well developed their "within" or consciousness is.

Psychic energy, the "within" and "without" of things, and the law of complexification—these are the three foundation stones upon which all the rest of Teilhard's thought relies.

We have observed the beginnings of the evolutionary process and its progression from simple, innumerable, minute, alike particles to ever more complex units. This complexification continues in both the crystallizing world (minerals) and the polymerizing world (nitrates, carbonates, hydrates). But only in the polymerizing world are true natural units to be found—units which are the result of a true combination or union—a number of molecules joining together to form a megamolecule which is something new, something more than the molecules which joined together to form it. It is in the polymerizing world that the evolutionary process eventually leads to life, called into life by the impulse of psychic energy, the light within (Book I, Chapter III, *The Phenomenon of Man*).

Part II: Life

When a certain stage of exterior and interior complexification had been reached and conditions on the earth were favorable, life was born, once and once only.

Although the evolutionary process is slow and gradual, there are critical stages or points before which something did not exist but after which it is found.

Once again I cite the homely example of how the pressure cooker works. When the temperature within the cooker reaches a certain degree, the atoms within the molecules of whatever stuff is being cooked, move faster and faster, expanding all the while. If the cook does not watch the gauge a critical point is reached and the stuff within explodes through the opening under the gauge and one has a mess all over the stove or even on the ceiling.

These "critical points" are points at which sudden changes occur, changes like the movement from non-life to life and eventually the movement from life to thought. The movement from this non-life to life is the transit from the world of the inanimate molecule to that of the virus and finally to the world of the cell, the living world. Just as the atom is the natural granule of simple elementary matter, so too is the cell the natural granule of life. (*The Human Phenomenon*, page 80).

The "without" of the cell represents the discovery of a new method of agglomerating a larger amount of matter in a single extremely complex unit. So too the "within" of the cell represents a new arrangement of the particles of consciousness—a new complex arrangement—a metamorphosis—not a totally new beginning. (The "light" simply grows brighter with the advent of life.)

Teilhard suggests that the milieu for the initial manifestations of life was a shoreless, warm ocean with few peaks of land—water- filled with free valencies—very unstable. As we get as near as we can to the threshold of life, it manifests itself to us simultaneously as microscopic and innumerable. At the critical stage that marked the transit to life, either there were a few cells that emerged at a few points that multiplied rapidly, or cells emerged at many points simultaneously (*The Phenomenon of Man*, pages 88-89 and 90-94).

Because so many conditions had to be present for life to emerge from non-life over a very long period of time, Teilhard maintains that the transit from non-life to life was a unique movement that happened *naturally,* that is, without the contrivances of humans in a laboratory, once and once only. He doesn't deny that it might be possible in the time after his writing, that humans might accomplish the transit from non-life to life in the laboratory artificially (*The Phenomenon of Man,* page 102).

His subsequent description of the expansion of life generally follows the Darwinian theses and hence needs no elaboration here. He devotes an entire chapter to this topic in *The Phenomenon of Man* that is explicit and detailed. (Book II, Chapter II, *The Phenomenon of Man*, Part II, Chapter II, *The Human Phenomenon*). It is only when he takes the next step in Chapter III, "Demeter," that he becomes original again when he comes to grips with the question of purpose or direction in evolution.

In his day, the first half of the twentieth century, the overwhelming majority of biologists would deny that evolution was *directed*, that it had a purpose, that it was going somewhere eventually. In our own day, the beginning years of the twenty-first century, while most biologists seem to feel the same way, there is a growing number who do not, thanks in part to the influence Teilhard's ideas have had on portions of the scientific community.

Teilhard states unequivocally that evolution is directed, it has a purpose, it is going somewhere. To him, evolution is simply the ascent to consciousness, which is marked by the continual growth (complexification) of psychic or radial energy over time. This development of the "within" is connected to and profoundly influenced by the growth (complexification) of the physical, the "without," the mechanical layer, tangential energy.

Arrangement, certain developments of the "without" make the "within" more and more perceptible. The selective mechanism of the brain and nervous system is the key that opens the door to higher and higher levels of consciousness, psychism of various levels. This is true of all living beings. We can even arrange them time wise along the ladder of evolution, that is, when individual living things *first* appeared, by the complexity of their nervous systems, especially their brains, their degree of "cerebralisation."

The deepening of consciousness reveals in each instance an increasing manifestation of the inner light—the light that in the human makes it possible to give meaning to his or her perceived world (*The Phenomenon of Man*, pages 142-144 and *The Human Phenomenon*, Chapter III).

Teilhard insists that in order to decide whether humankind is superior to the rest of the animal kingdom, we have to push aside all what he terms "secondary" and "equivocal" forms of inner activity in the human *phenomenon of reflection*. (Because of the light "within," that is, psychic energy, to be able to think in a way similar to how God, "The Light," thinks). It is important to settle this question of "superiority" of the human in terms of reflection because pure knowledge, the basic way we perceive our world, the meaning we give to it, the ethics of life, depend on how we answer the question: Is the way a human knows (reflection) superior to the way all other living things "know?"

Reflection is experimentally defined as the power to turn in upon itself as an observer looking at an external object— only the object is the actual thinking or behavior of the *thinker*. No longer simply to know, simply to experience, but also to know (be aware of) that one knows. Animals know, but there is no reliable evidence that an animal has self-knowledge, that it knows that it knows! If it could, animals would be the inventors. Animals would have developed a series of internal constructions, ways of thinking, logic etcetera., which could not have escaped our systematic observation. (*The Phenomenon of Man*, pages 165-66).

Thus, the essential difference between the psychism of man and that of animals is that only man knows that he knows. Are the animals therefore all alike in their psychic makeup? Teilhard answers with a resolute "no." He says that there are many divergent opinions concerning the psychic makeup of animals. He cites three representative ones, each of which has an element of truth, but also a cause of error:

1. The Scholastics who viewed instinct as a sort of subintelligence, homogeneous and fixed, marking one of the onto-logical and logical stages by which being grades downwards from pure spirit to pure materiality;

2. The Cartesians who maintained that only thought existed, so the animal, devoid of any "within," was a mere automaton; and

3. Most modern biologists who draw no sharp line between instinct and thought—neither being very much more than a sort of luminescence, "an emission of light occurring at a temperature below that of incandescent bodies" (*Random House Webster's College Dictionary*, 1999 edition, page 792) (a shining without heat—a form of the inner light) enveloping the play—the only essential thing—of the determinisms of matter.

Teilhard then proceeds to describe in detail how the nervous system becomes more complex and the brain larger and convoluted as we move up the evolutionary ladder of the biota until we reach mammals then the primates, and finally man, who has the largest brain, the best-developed nervous system, and therefore the highest and best-developed awareness or consciousness—(capable because of the light "within" to see, that is, understand in greater depth, the light "without").

In addition to the nervous system and the brain, there are certain other "withouts" which aid in the development of the "within," such as a larger cranial capacity which makes a larger brain possible, eyes in front of the head, and being biped which freed the hands to lift objects before the eyes for examination.

Certain other outer structures effectively limited the development of the "within," such as the chitinous skeletons of insects which keep insects too small to develop large enough brains, and also too specialized structures, such as teeth, jaws, etcetera, which allow an animal to survive without reliance on its brain.

Part III: Thought

It is precisely because the primates have relatively weak bodies that nature went to work on the brain as the instrument of their survival—and the brain is directly related to the complexity of the "within" (*The Phenomenon of Man, pages147*-160). When the complexification of the "without" (brain and nervous system) and the "within" of a certain line of evolution (the mammals) had reached another critical point—thought was born in one phylum—the hominids—and man was born.

When Teilhard was writing in the first half of the twentieth century and even now at the beginning of the twenty-first century biologists are not yet agreed on whether or not there is a purpose or end point for evolution. Nor is there majority agreement among psychologists as to whether the human psychism differs specifically (by "nature") from that of other animals. In fact, in 2015 as this book is being revised, we see the drastic use of animal "humanizing" in pet cemeteries, day care and play dates for animals, animal psychiatrists, and a number of persons who maintain that animals have immortal spirits also. In the minds of many, there is no line at all, (or at most a very fine and not too important a line), between animal and human intelligence and rights.

Teilhard asserts that instinct is a phenomenon that through its many expressions reveals the very phenomenon of life in its many dimensions. There are many kinds of animal behavior, and many forms of instincts. There is the psychical makeup of an insect, the instinct of a squirrel, that of a cat or an elephant, each different according to its position on the tree of life. These varied and complex instincts constitute a growing system, especially in the higher animals (particularly, in the great apes), whose behaviors recall those behaviors of which we make use to define the nature, and prove the presence in ourselves of "a reasonable soul." If the story of life is no more than a movement of consciousness veiled by morphology (the form and structure of organisms), it is inevitable that, as the evolutionary ladder approaches man, the psychical makeups of animals seem to reach the *borders of intelligence* (*Random House Webster's College Dictionary*, page 863).

Though Teilhard died before much of the modern experimental work was done with chimpanzees, dolphins, and whales, he would not have been surprised at the findings. But though animal instinct reaches the borders of intelligence (by which he means reflection), the movement from instinct to reflection was a leap that he characterizes as something altogether new that emerges at a critical point after a long period of innumerable small changes. Because the changes are so small and take so long, it is not realistic to expect to find an intermediate state between instinct and thought. There is a sudden leap from instinct to thought, a leap that presupposes the psychical transcendence of thought over instinct, or we have to accept that individual "A" operated from pure instinct and then, when conditions and changes were exactly right, individual "B" appears with the power of reflection. The transition to reflection (like that from nonliving to living) involved a change of state and then the beginning of another kind of life—the interior life.

"Something" has become "someone." The grains of matter and life refer to "something." The grain of "thought" to "someone" (*The Phenomenon of Man*, pages 167-168, 170-71, and 173.

To describe this process or transition from instinct to thought (reflection), Teilhard coined the word "hominisation." The process of hominization is both individual and collective. All the psychisms of all the related group of organisms called "animals" including man, are part of the developing spiritualization, the increasing and deepening levels of consciousness as evolution progresses (*The Phenomenon of Man*, page 180), (growing in the light and toward "The Light").

Thus "hominisation" can also apply to the continuing process by which man becomes more human and this entire "humanizing" process is intimately bound up with the development of the mind, of the power of reflection. This development Teilhard calls "noogenesis." When the last creature of instinct knew that he knew, this creature added reflection to his "within" while still retaining some instinctual powers. When this occurred, all of civilization took a step forward which was really a gigantic leap! Teilhard cites the conventional geological concept of the zonal composition of our planet. The barysphere or core central and metallic, the rocky lithosphere, the fluid layers of hydrosphere and atmosphere, and the biosphere of animals (fauna) and plants (flora).

But then he adds a new layer that began with the advent of man— the thinking layer— which has gradually spread over the entire world largely because of the spherical shape of our globe. This layer he calls the "noosphere" (*The Phenomenon of Man*, page 182).

It cannot be emphasized too strongly that to Teilhard the noosphere or thinking layer is as real a layer as any of the other layers that make up our world. In fact, the advent of this most recent of layers marks a critical change in the evolutionary process. Up to the point where life took the critical leap to thought, evolution proceeded by chance and groping. But with the advent of the noosphere, the thinking being, man, begins to take charge of evolution. Evolution becomes centered in a developing consciousness. This is where future significant change will take place. Evolution of the "without" will continue, but most of whatever further physical change there is will be the result of thought—planned change—rather than change through chance and groping.

The beginnings of man have always been a source of conflict between men of science and men of religion. To Teilhard this conflict is not a necessary one.

There was no big bang, no perceivable cosmic upheaval when man appears on the evolutionary scene. Archeological digs have revealed myriads of stone instruments from various parts of the world found in scattered piles with evidence of fires and some sort of language(s) indicated by the earliest of cave drawings. So science sees man as coming into the world as a group slowly over many thousand years.

However, this does not mean that all humans could not have been descended from a common pair (Adam and Eve). It only means that there are next to zero odds that there would ever be found any fossil remains of a single pair of humans. So there is no scientific evidence to support descent from an original pair of humans.

Therefore, science cannot speak authentically of monogenism, that all humans had their origin in a unique human pair. But science can, based on the evidence cited, speak decisively in favor of monophyletism, that all humans are descended from a single phylum (*The Phenomenon of Man*, pages 186-88).

Teilhard spends the rest of Part III, Chapter Two, in presenting the evidence for his contention that men are descended from a single phylum. Teilhard's position on monophyletism does not negate the Apostolic Christian teaching which while allowing for the evolution of the body, maintains that God, The Light, directly creates *each* human soul (spirit) that animates the body and makes it a human body.

As the *Catechism of the Catholic Church* states (p. 93, par. 365):

The unity of soul and body is so profound that one has to consider the soul to be the "form" of the body: i.e., it is because of its spiritual soul that the body made of matter becomes a living, human body; spirit (the within) and matter (the without), in man, are not two natures united, but rather their union forms a single nature (human nature).

Teilhard also deals with the question of why man remained one species rather than differentiating into a number of species, as have other living things. The basic reason he cites is the tendency of the conqueror to assimilate with the remnant of the conquered through intermarriage. (*The Phenomenon of Man*, page 209).

Of course, as modern discoveries have shown us, some groups of men remained isolated—but they are a tiny minority of the human species. The majority of men lived in regions that are more favorable to "the concourse and mixing of races—extended archipelagoes, junctions of valleys, vast cultivable plains, particularly, irrigated by a great river." Teilhard lists five concentrations of men that had a particularly important role in building up the noosphere:

1. the Indian Civilizations of Central America;

2. the Polynesian Cultures of the South Seas;

3. the Chinese Civilization Along the Yellow River;

4. the Indian Civilization Along the Ganges River;

5. the Egyptian Civilization Along the Nile Valley; and

6. the Sumerian Civilization in Mesopotamia.

The interaction over time of these civilizations/cultures gave the "within," the "psychosomatic currents," a continuing push forward (*The Phenomenon of Man*, page 210).

Teilhard examines each of these centers of civilization in turn and concludes that, "the facts force us to recognize that during historic time the principal axis of anthropogenesis, the development of the human phenomenon, has passed through the west".

This happened for a number of reasons which Teilhard cites in detail, but which in general can be attributed to lack of isolation in the west, and hence a constant intermingling of various groups of people, and also a double concentration on metaphysics and physics on the "within" and the "without," which impelled western civilization to undertake the building up of the world, not to abandon it, or to consider the "without" an illusion and only the "within" as truly real.

The basic proof that the principal development of the human phenomenon has passed through the west lies in the fact that from one end of the world to the other, all the peoples, to remain human or to become more so, are inexorably led to formulate the hopes and problems of the modern earth in the very same terms in which the West has formulated them (*The Phenomenon of Man*, page 212).

Although man from the very beginning began to take charge of evolution through his inventions, and later through controlled breeding of plants and animals, it is only in the last hundred years that mankind generally perceived the reality of evolution. The dawning of the modern earth began with the Renaissance. Teilhard cites three changes that mark the birth of this new age:

1. Economic changes—from a civilization based on the soil and its participation to a money economy. Ever increasing portions of populations became isolated from the land as they massed into larger and larger towns and cities hoping for a better, that is, more affluent and less arduous life.

2. Industrial changes from one known source of chemical energy: fire, to coal, oil, and gas—from only one sort of mechanical energy: muscle, human and animal, to the many "horse-power" machines. More and more people lived in communities of factory chimneys and offices.

3. Social changes and the awakening of the masses. The agglomeration of people into larger and larger groups living in extremely close proximity to each other and working under often barbaric conditions led to new social groupings like labor unions, increases in numbers of fraternal organizations, and new political movements like Marxism.

Teilhard is restating his basic premises of the unity of all that is, and of the gradual building up of the universe from the simple to the complex, from the inorganic to the organic, from the world of chemistry to the world of biology, and from the biosphere to the noosphere—from instinctual life to reflective life. One cannot understand any phase of this process without understanding what preceded in this phase, and what has or may follow it. This entire all-encompassing process is what he terms "evolution." To those who can use their eyes, nothing escapes evolution—the mind, consciousness, has evolved with the body. Evolution is now focusing more and more on the psychic zones, the "within," and not treating the "within" as secondary to the "without," but rather making the "within" the driving force, the very heart of the "without!"

Bernard J. Fleury Ed.D.

The exterior world, the "without," in the process of becoming (cosmogenesis), reached up to mind and now confronts us with an interior world, the "within," the world of reflective thought in the process of becoming (noogenesis).

Beneath all these physical changes in the "without" of things, beneath this change of age, and partly at least because of it, there was a change of thought, a change of soul, an awakening to new ways of looking at the universe and man's role in it. People began to see things in a different light. Life assumed new meanings as the inner light grew brighter.

Very slowly but surely, humans began to realize what Teilhard terms as the "irreversible coherence" of all that exists. We have been forced to look at ourselves, who we are, what we do to ourselves and to our environment, what we want out of life over time and space, recognizing that time and space are joined together, and this joining weaves the energy of the universe together. Truly modern man, according to Teilhard (and many are not yet "modern"), is becoming capable of seeing things in terms of space, time, *and* duration—what he calls "biological space-time," and becoming willing and incapable of seeing *anything* otherwise, not *even himself* (*The Phenomenon of Man*, pages 215-219 and 221).

The evolutionary process is a continuous movement from the simple to the complex, and cosmogenesis (the world in the process of becoming) incorporates within itself a developing consciousness that becomes more and more advanced, until it achieves the complexity of reflective thought—a noogenesis (reflective thought or minds in the continuous process of further complexification and deepening awareness).

Because man has the most highly developed consciousness, and, as we have seen, the most highly developed consciousness is the growing edge of evolution, Teilhard characterizes man not as: "The center of the universe but something much finer—the arrow pointing the way to the final unification of the world in terms of life". This is true because the human being (man) is the newborn, most complexified of all the layers of life that have succeeded each other in the ages long process of evolution. In the beginning was The Fire, Alpha, The Light, and at the end is Omega, The Light (*The Phenomenon of Man*, page 224 and *The Human Phenomenon*, page 156).

Part IV: Superlife

Teilhard begins his thoughts on Superlife or the Future by asking whether in all our theories on evolution and the ways we have applied these theories, we have emphasized the mechanical and impersonal at the expense of the place due to "person" and the strength, the effect, the influence of personalization in the movement and shape of the evolutionary process.

As we have already seen earlier in Teilhard's vision of the goal or purpose of evolution, that evolution is an ascent to consciousness (reflection). This is an ascent from the simple to the increasingly complex with consciousness being the most complex interior expression of a highly complex "without" or physical structure, that is, the brain and nervous system. Since this has been the direction in which evolution has progressed to this point, and since scientific prediction for the future is based on repeated and sustained past behaviors, it seems logical to predict that evolution's fulfillment will be in some type of supreme (even more complex) consciousness (The *Phenomenon of Man*, page 258).

This supreme consciousness Teilhard calls "Omega" because this consciousness is the goal or end of evolution. Omega is personal. This is logical since personalization is simply an ever-deepening consciousness—the more highly developed the consciousness, the more we can apply the term "person" to each centered unit of such consciousness.

Also, as we have seen, evolution moves from the simple to the complex, from non-life to life, from life to thought and consciousness, and thence to deepening consciousness.

Thus, if Omega is the goal of evolution, Omega must be personal, must be a supreme consciousness, a supreme person, for the personal is attributed a higher (more complex) order of existence (humankind)—and how can a higher order of existence (humankind) be capped, crowned, or have as its goal the impersonal, a lower order of existence?

Union with Omega is not a union with a kind of world spirit into which all lesser spirits will be submerged. Person is an immortal characteristic of humankind, for Teilhard does not subscribe to the concept of God as one who reabsorbs into himself the spirits of individual humans once their earth lives are ended. Omega (God, the end or goal of all things) is not a grand stew of elements that have had a temporary earthly existence and are now part of the stew. Rather, God is indeed the magnetic center whose force draws all other individual centers back to himself—to be reattached to him in an intimate, personal, eternal union, but not to a union which annihilates the individual center (person). Humans become most fully human only when they are finally grafted onto Omega, and He becomes all in all to them (*The Phenomenon of Man*, pages 262-263).

It should be noted here that Teilhard speaks of Omega point as the focus of union, while Omega is the Supreme Consciousness that is the center of the union. The relationship between Omega and individual centers of consciousness may be likened to the relationship between the planets of our solar system and their sun. The energy and attraction of the sun both gives life to its planets and holds them in an orbit about itself. Yet each of these planets remains a distinct entity—a center of existence.

Omega is in all things by the very fact that the basic stuff of the universe is psychic energy (light) and Omega is the Alpha, the source of this energy (and is Himself Psychic Energy, "The Light," centered in one Divine Nature (Being) of three Divine Persons). Alpha and Omega, the first and last letters of the Greek Alphabet, are often used to indicate that God is the beginning and end of all that exists.

But *all things are not Omega* for Omega is transcendent, above all creation. Creation is part of Omega, but Omega is not part of creation. Omega is actuality (there is no potential, no change—no evolution in Omega), autonomous (dependent on no one or no other thing), irreversible (again indicates no element of change or disintegration). In Book IV, Chapter 2, Teilhard gives detailed explanations for listing the preceding attributes as essential to Omega.

We have seen that the future of man is intimately bound up with the future of all men if mankind as a whole is to move toward Omega point. Forced collectivization is not the answer, but voluntary drawing together is. What is to be the source of such a voluntary drawing together? Teilhard answers with the word "Love," the affinity of being with being. "Love" is a grossly overused word today with many and contradictory meanings, so to understand Teilhard's use of the term we should consider some examples of "love" which he lists: sexual passion, parental instinct, social solidarity, etc. It is a "real internal propensity to unite," present at least in rudimentary form in the simplest molecule. This uniting, however, is a union of center to center of what is deepest and most vital in each partner.

It is not difficult to understand "love" on a one-to-one or small group basis; but when we begin to speak of universal love—all men loving everyone and everything—our reason hesitates. Teilhard perceives this hesitancy when he states that the capacity to love intensely seems limited to one human being or to only a few who are very close to us by blood or affection. To say that one loves everything and everybody is a phony sentiment, not a love that represents a strong affinity of being with being.

But, despite what we have just written, we also see evidences of a universal love—a love for the "all" when we are stirred by a sense of the grandness of the universe, the beauty and wonder of nature, music that stirs the hearts of all assembled to hear it—that draws them together as one voice in singing the lyrics or sharing in one feeling of patriotism, love of God, and of others. There seems to be a "Great Presence," a magnetic personal force drawing us to a universal love—a love in the "all." When we begin to see ourselves as elements of a developing cosmos—begin to develop a communal cosmic sense—we come to realize that such a universal love is not only possible but is the final way that we as individual centers of consciousness are called and able to love.

If indeed we are called to a universal love, how do we account (especially since September 11, 2001) for the appearance worldwide of increasing terrorism and obscene manifestations of hatred? Just this, that we must place people above ideologies, political systems, land, power, and money. A note of caution must be inserted here. We must remind ourselves of duration—the slow pace at which cosmic change takes place. It has only been a couple hundred years since "modern man" has had a glimpse of a higher state of union while still in the struggle of groups of humans to be free. It is foolishness to think that the earth will be transformed before our eyes in the space of a generation or two since it took from a half million to a million years for the pre-hominids to evolve to modern humans. There must be some object or source of love who is universal and "above our heads" (*The Human Phenomenon*, pages 180-181, and *The Phenomenon of Man*, pages 265- 267).

Thus, if Omega is to be lovable, it must be someone rather than something. Love demands a beloved—not an abstract notion of a person but a real flesh-and-blood being to whom one may relate— in this case a real person to whom all humans can relate. It is at this point that Teilhard's

Christianity becomes the source of his reasoning, for this real person who can serve as the love object is none other than Jesus Christ—true God and true man. Teilhard's Christ is an integral part of the evolutionary process. All evolution before Christ was a preparation for his coming. All development since His coming is simply the completion of His mystical body. Christ provides the basic elements of hope and love. Hope, as we have seen earlier, is essential to progress and love is essential to unification and convergence that are also essential to progress. Christ is the love object who will draw all humans together.

In Chapter 4, "My Universe" of his book, *Science and Christ,* Teilhard speaks of Christ's role in three dimensions:

1. The revealed Christ is identical with Omega.

2. It is inasmuch as He is Omega, that He is seen to be attain- able and *inevitably* present in all things.

3. Finally it was in order that He might become Omega that it was necessary for Him, through the Travail of his Incarnation, to conquer and animate the universe.

The first point simply refers to the concept of Christ being all in all—the focal point—the goal of evolution. In the second point Christ is presented as the attainable and universal element. How is Christ the "attainable" and "universal element?" In his epistles, St. Paul writes that the world was created (in whatever manner) by God the Father, through His Word (God the Son), and *for* His Word. Thus, Christ is the element of syntheses, the unique Spirit who holds all the elements of the cosmos together at their ascending (evolutionary) order of being (their complexity, their consciousness). He is the unique "Monad" which gives everything below Himself its meaning, its "power of action and reaction."

Since Christ is Omega, His organizing and energizing

activity affects all areas of our being both the "within" (spirit) and the "without" (body). His power extends to all that is human both individual "withouts" and "withins" as well as the overall process of cosmic evolution. He is eternally present as we humans, as St. Paul writes, "groan in pain as we await the redemption of our bodies."

Every good human action, every earth process that leads to a unification born of love, is taken up by Christ who "transforms and divinizes it" (*Science and Christ*, pages 54, 57, and 59).

This is no plaster statue Christ who is relegated to a pedestal on a church wall! Teilhard's third point reinforces the vital role of Christ in animating the world.

Since according to St. Paul, the world was created through Christ and for Christ, His influence on evolution exists from the very onset of creation. All of history before His conception in the womb of Mary was a preparation for His coming. It took the struggles of primitive humans, the civilizations of Egypt, Greece, and Rome as well as Israel's call and longing, for the world to be ready for the bud (Christ) to bloom on the stock of Jesse and humankind. All these developments in thought, customs, physical structures (like the labyrinth of roads constructed by the Romans) were necessary to make the world ready for Christ's incarnation. When Christ was born on that first Christmas, a new era dawned for the world (*Science and Christ*, page 61).

Such a universal cosmic view of Christ will lead the believer to a conquest of the world.

Teilhard's vision of the Cosmic Christ was based on the historical Christ as seen in the Gospels and particularly in the writings of the Apostle Paul. Teilhard also had a special devotion to Christ under the image of Christ's Sacred Heart. The connections to light are pervasive and profound.

In a letter he wrote from the front in World War I, where he served as a stretcher-bearer, Teilhard writes that the vision of the Sacred Heart of Jesus given to St. Margaret Mary was particularly meaningful to him. In it she saw herself as an "indeterminate atom striving to lose herself" (by being drawn) "in the great center of light that was the heart of our Lord." Teilhard saw two elements that summed up life for him; first, his "absolute dependence on the creative and sanctifying energy of God," and second, "the passion for God" that maintained for Teilhard a passion for life. This entering of God into his whole achievement, made everything he did a union and cooperation with a self-giving and transforming God.

In 1923, Teilhard recorded another meditation on the Sacred Heart in which he relates that the visions given to St. Margaret Mary reveal a Cosmic Christ, a Christ whose influence and image are vastly greater than the limited image we had before this vision. The image of Christ that St. Margaret Mary saw shows the heart of Jesus exposed, a furnace of fire that draws one's attention to it so much so that the rest of Christ's image seems to fade away and all that Teilhard can see in Jesus is "the face of a world which has burst into flame" (*The Spirituality of Teilhard de Chardin*, pages 66-67).

But it is important that we should not interpret this and similar passages in terms of a vague mystagogy. As a scientist, Teilhard naturally started from the facts given in the investigation of some individual concrete situation, even though his speculations soared far beyond that particular event. So it is with his religious thought.

Though Teilhard gives a much grander and more universal place to Christ in all aspects of evolution, he has no intention of suppressing the historical reality of Christ. Just the opposite! He writes that the universal and mystical Christ that St. Paul first described is simply an "expansion of the Christ who was born of Mary and who died on the cross....However far we may be drawn into the divine spaces opened to us by Christian mysticism, we never depart from the Jesus of the Gospels" (*The Spirituality of Teilhard de Chardin*, page 68).

But we must be open to an ever more in-depth understanding of the more universal implications of Christ beyond the space-time of his life on earth or risk forgetting that the totality of his teachings, and its implications for all time, had to be progressively unfolded and appreciated under the influence of his spirit as the evolution of the "within" progressed.

There have been tremendous changes in man's vision of the universe since the years of Christ's earth life, and this has had an impact on religious practice and belief.

The discoveries of science that reveal the vastness of the universe, seem to make the human a tiny speck, a small part of the universe that is much more beautiful and important than the human being.

We have also learned much more about the size and unity of the world "within" us with the new ideas coming from psychology and sociology.

So, with all these new insights, is the Christ of the Gospels, whose earth life was confined to a small part of the Mediterranean world, still capable of embracing and "forming the center of our prodigiously expanded universe?" Is the Christ of the New Testament irrelevant in the modern world?

Teilhard's answer is St. Paul's that was written in the first generation after Christ's resurrection. Speaking of Christ, he has this to say: "He is the image of the invisible God," his is the primacy over all created things. "In him everything in heaven and on earth was created"…the whole universe has been created through him and for him. He is at the head and front of all things; in him all things are bound together. He is also the head of the body, the Church; its origin as firstborn of the dead. You see he was to be supreme in all things. For God chose to establish in him the fullness of all that exists. Through him God chose to reconcile the whole of reality in him reconciling through him alone everything that is on earth or in heaven (1 Colossians: Verses 15-20).

Thus, from its very beginnings, Apostolic Christianity has proposed that in the being of the historical Jesus of Nazareth, the whole explanation and purpose of all of creation is to be found. The answer to the riddle of the universe is: Christ Jesus.

But if this answer is to satisfy modern man, we have to update our Christology to reflect the cosmic picture modern man will support. Christ needs to be presented in terms of his cosmic function as indicated by St. Paul (in 1 Colossians previously cited), a Christ at the very heart of the universe, not dissociated from it as a detached (and irrelevant) fragment.

Our new Christology requires that we conceive Christ to be constituted as the cosmic center of creation. Simply as a magnification, a transformation, realized in the humanity of Christ of the *aura* that surrounds every human monad, Light from "The Light" (*The Spirituality of Teilhard de Chardin*, pages 71-72).

In the case of other human beings, we have experience, from time to time, of leaders who are at once centers of attraction and points at which some corporate activity concentrates and develops. We can think of great political figures, such as Abraham Lincoln or Winston Churchill; we can think of great artists, such as Michelangelo, Shakespeare, or Beethoven; we can think of the great inventors and scientific thinkers, like Copernicus, Lister, Marconi, and Rutherford. All these men have given a fresh impulse to the human spirit by providing centers of psychic energy, around which the efforts of countless individuals have collected and found inspiration. Analogously, we can think of Christ, the man, the son of Mary, chosen so that his influence, his *aura,* might serve as the medium in which men's united efforts could be enlarged and given a fresh direction dominating and drawing them all to himself (*"Forma Christi"*, Writings in Time of War, page 253).

There is only *one single center* in the universe; it is at once natural and supernatural; it impels the whole creation along one and the same line, first towards the fullest development of consciousness, and later towards the highest degree of holiness: in other words, towards Christ Jesus, personal and cosmic. If the universe is constituted of its myriad elements, the elements in turn need the universe for their existence. So we must say of every man that, over and above his body and soul, there is a physical relationship binding him to the universe in which he finds his fulfillment. And, just as he himself is in a continuous process of development, of becoming himself, so is the whole cosmos a growing and developing whole. Just as the cells combine to make a human body so do human beings themselves combine to make up that total, but yet unrealized, fulfillment which is the cosmos-to-be.

In his human nature, Christ is this individual, born of Mary, nurtured in Bethlehem and Nazareth, "advancing in wisdom" as he learned from his mother, his teachers, his friends. He too needed the support of nature's resources, and of friendship and love. At that level, he is an element in the cosmos. But we know, from St. John and St. Paul, that he is also the center of creation; the force which can subject all things to itself the origin and the term of the whole cosmic process—Alpha and Omega. Hence, just as there is a certain physical relationship that both perfects him and enables him to contribute to the perfection of the universe, so there is a relationship of all beings, but, in a special sense, of the members of his body the Church to Christ. Just as there is something of the cosmos in each individual constituent of it, so there is something of Christ in every creature. It is the responsibility and the privilege of each person to develop that

Christ-element within himself, thus adding to the fullness that Christ, in himself, already is *The Spirituality of Teilhard de Chardin*, pages 73-74).

The aura, the light that surrounds and is within each human being has as its source Jesus Christ, The Light. He is the Word of the Father, God from God, Light from Light.

When we examine the elements of the near-death experiences of countless persons in e-Book 4 of the Called Into Life by the Light Series we will see how one of the remembered elements is the aura of light that surrounds each person. A second common element is the profound experience of an all- enveloping "love" that is experienced as an external as well as an internal phenomenon. Teilhard's vision of God as "The Light" *and* Love seems to be borne out in these experiences.

Bernard J. Fleury Ed.D.

Teilhard de Chardin: Essential Points On Man, The Creature Who Mirrors The Light

1. Basic stuff of the Universe is Energy. ("Luminous in nature")

    a. Basic Energy is Psychic.

2. There is a "Within" (the Inner Light, every form of Psychism) and a "Without" (Matter, the Physical, A Body, Mass) of things – this quality of the earth was present from the very beginning.

3. Complexification: the stuff of the universe tends toward increased complexity as evolution progresses.

    a. There is a direct and positive relationship between the complexity of the "Without" of things and how well developed their "Within" or consciousness is.

4. When a certain stage of exterior and interior complexification had been reached and conditions on the earth were favorable, life was born, once and once only.

5. Complexification Continues.

    a. Certain outer structures limit the development of the "Within"
        (1) Chitinous skeletons
        (2) Small cranial capacity
        (3) Too specialized structures - teeth, jaws, etcetera.

    b. Certain outer structures aid in the development of the "Within"
        (1) Refined nervous system
        (2) Larger cranial capacity and hence a larger brain
        (3) Eyes in front of head
        (4) Biped - leaving hands free to examine objects

6. Birth of Thought

    a. When the complexification of the "Without" and the "Within" of a certain line of evolution (Mammals) had reached another critical point, thought was born in one phylum, hominids, and man was born.

    b. Man is distinguished from other conscious beings by reflective thought – knowing that he knows. The inner light appears.

    c. There are varying levels of consciousness short of reflective thought, in all living things.

7. Noosphere

    a. With the birth of thought a new sphere enveloped the earth – a sphere of thought or "noosphere".

    b. With the advent of this sphere, the thinking being, man, begins to take charge of evolution.

    c. Evolution is now centered in a developing consciousness – this is where future significant change will take place. This does not, however, rule out further physical change; but whatever change there is will be the result of thought – planned change rather than change by chance and groping.

8. Omega Point

    a. Evolution has a direction – change and process are directed.

    b. The crowning point of evolution is supreme consciousness (The Light) whose magnetic force will draw other highly developed "centres of consciousness" toward it. The relationship between

Omega and individual centres of consciousness may be likened to the relationship between the planets of our solar system and their sun. The energy and attraction of the sun both gives life to its planets and holds them in an orbit about itself. Yet each of these planets remains a distinct entity – a centre of existence.

c. Omega is in all things but all things are not Omega. Omega is actuality (there is no potential, no change), autonomy dependent on no one or no other things), and transcendent above all of creation (creation is part of Omega, but Omega is not part of creation). Omega is The Ultimate Light.

d. Omega is personal. This is derived from a value judgment, since evolution is moving in the direction of increased awareness, increased personalization, it appears absurd to assert that the impersonal will crown the personal, for the personal is attributed a higher (more complex) order of existence, and how can a higher order of existence be capped, crowned or have as its goal a lower order of existence?

e. Omega is the Godhead of which Christ (Light from Light), is one Person.

f. Christ plays an indispensable part in the evolutionary process. All evolution before Christ was a preparation for His coming. All development since His coming is simply the completion of His Mystical body. Christ provides the basic elements of hope and love. Hope is essential to progress and love is essential to convergence and unity, which are also essential to progress. Christ is the love object who will draw all men together. (Points "e" and "f" are dealt with in detail in de Chardin's *Science and Christ.)*

9. Directions and Conditions of the Future "Along what lines of advance, among others – judging from the present

condition of the noosphere – are we destined to proceed from the planetary level of psyche totalisation and evolutionary upsurge we are now approaching?" "I can distinguish three principal ones in which we see again the predictions to which we were already led by our analysis of the ideas of science and humanity. They are the organisation of research, the concentration of research, upon the subject of man, and the conjunction of science and religion. These are three natural terms of one and same progression." (Book 4, Chapter III, *Phenomenon of Man)*, and Part IV, Chapter III, of *The Human Phenomenon.*

10. Man is both free and determined. "Reflective substance requires reflective treatment. If there is a future for mankind, it can only be imagined in terms of harmonious conciliation of what is free with what is planned and totalized." (Book 4, Chapter III, *Phenomenon of Man,* and Part IV, Chapter III, of *The Human Phenomenon.*

Bernard J. Fleury Ed.D.

Summary

--Teilhard de Chardin: The Man Who Christianized Purpose Driven Evolution.

He was a man who had the capacity to see all as part of one, to combine a basic scientific theory of evolution with a philosophical and theological vision. It is a light based theory.

--Prologue: Seeing

Teilhard asserts that seeing correctly is essential if we are to understand the universe as it is. In order to see correctly, we have to see ourselves, the human being, as the "center of perspective."

--The First Basic Point of Teilhard's Process of Evolution

There is a "within" (every form of psychism-consciousness) and a "without" of things (external structure – brain, bones, nervous system) – this quality of the earth was present from the very beginning.

--The Body (Mass) Called Light

Newton and other mathematicians moved toward a material atomism (Galileo) that posited the earth as a grand mass that was the sum of numerous tiny masses.

--Was Descartes's Dream – A Vision or an Illusion?

Descartes saw space or "extension" as filled with Matter – an "atomistically conceived plenum" (space) along which any action between the eye and the object perceived, travels.

--List of Persons and Their Thoughts/Findings on Light in the Seventeenth through Twentieth Centuries.

L. Euler: Luminous objects vibrate.
R. Boyle: Air not necessary for transmission of light.
T. Young: Vibratory theory of light.
A. Fresnel: Developed wave theory of light.
F. Arago: Confirmed results of Fresnel's experiment.

--The Light of Electricity: A way of Seeing.

Michael Faraday, is one of the greatest experimental scientists of all time. In nature, he saw nature's God, so his theology and philosophical science were two sides of a single reality for him.

--Light and Magnetism.

James Clerk Maxwell was the discoverer of the laws of electrodynamics. Like Teilhard de Chardin, Maxwell was a synthesizer.

--The Second Basic Point of Teilhard's Process of Evolution: The Law of Complexification.

The stuff of the universe tends toward increased complexity as evolution progresses, and there is a direct and positive relationship between the complexity of the "without" of things and how well developed their "within" or consciousness is.

--Introduction to Teilhard's Purpose Driven Evolution.

Teilhard de Chardin was a great synthesizer. Science and Religion, Faith and Reason and Spiritual worlds had to be seen as two aspects of the evolution of the entire universe including humankind.

The process is Purpose Driven evolution. All of creation is Purpose Driven. It begins with Alpha, The Light (God) and will come to an end or fulfillment at Omega Point – final union with God (Omega) our beginning and our last end.

--Light: Foundational Energy of the Universe.

This interpretation of Teilhard's Purpose Driven Evolution from the Human Phenomenon by Sarah Appleton Weber is categorized under four basic parts:
1. Prelife
2. Life
3. Thought and
4. Superlife

--Part I: Prelife

What is the basic stuff of the Universe? What is the

ultimate or real? Teilhard begins his answer to these questions by writing, "Elementary Matter...the stuff of tangible things "that" reveals itself with increasing assistance as radically particulate, yet basically connected, and finally, prodigiously active. Plurality, unity, energy: The three aspects of matter." (*The Human Phenomenon*, Page 12).

--Part II: Life
When a certain stage of exterior and interior complexification had been reached and conditions on the earth were favorable, life was born, once and once only. Although the evolutionary process is slow and gradual, there are critical stages or points before which something did not exist, but after which it is found.

--Part III: Thought
It is precisely because the primates have relatively weak bodies that nature went to work on the brain as the instrument of their survival – and the brain is directly related to the complexity of the "within" (*The Phenomenon of Man*, Page 147-160). When the complexification of the "without" (brain and nervous system) and the "within" of a certain line of evolution (the mammals) had reached another critical point – thought was born in one phylum – the hominids – and man was born.

--Part IV: Superlife
Superlife begins Teilhard's thoughts on Superlife or the Future by asking whether in all our theories in evolution and the ways we have applied these theories, we have emphasized the mechanical and impersonal at the expense of the place due to "person" and the strength, the effect, the influences of personalization in the movement and shape of the evolutionary process.

--Teilhard de Chardin: Essential Points on Man, The Creature, Who Mirrors The Light.

His ten Points:

1. Basic Stuff
2. Within and Without
3. Complexification
4. Life
5. Complexification Continues
6. Birth of Thought
7. Noosphere
8. Omega Point
9. Directions and Conditions of the Future
10. Man is both free and determined

Bernard J. Fleury Ed.D.

Bibliography

Corbishley, Thomas. 1971. *The Spirituality of Pierre Teilhard de Chardin.* Paramus, New Jersey: Paulist Press.

De Chardin, Pierre Teilhard. 1965 *"Forma Christi".* Writings in Time of War Now available as part of Pierre Teilhard de Chardin (Modern Spiritual Master Series) by P.T. de Chardin and Ursula King. May 1, 1999. Maryknoll, New York: Orbis Books.

—. 1965b. *Science and Christ.* New York. Harper and Row.

—. 1965c. *The Phenomenon of Man.* New York: Harper & Row.

—. 1999c. *The Human Phenomenon.* Edited and Translated by Sarah Appleton – Weber. Portland, Oregon: Sussex Academic Press.

Wikipedia Contributor. Accessed July 17, 2015. René Descartes, March 31, 1596 to February 11, 1650.
—. Isaac Newton, January 4, 1643 to March 31, 1727.
—. Robert Boyle, January 25, 1627 to December 31, 1691.
—. Leonard Euler, April 16, 1707 to September 18, 1783.
—. Thomas Young, June 13, 1773 to May 10, 1829.
—. Dominique Francois Argo, February 26, 1786 to October 2, 1853.
—. Augustine–Jean Fresnel, May 10, 1788 to July 14, 1827.
—. Michael Faraday, September 23, 1791 to August 25, 1867.
Author Bernard J. Fleury's Note:
I had learned about some of Faraday's experiments in my discussion with engineer David Tumey as cited in Part Two: Conversations on Royal Rife Ray Tube in my e-Book *The Mind-Body-Spirit Connection In The Medicine Of Light.* He was talking about a double-walled Faraday cage he and his partner William Sheline had acquired.

Teilhard de Chardin The Search For The Light In Evolution

I researched Michael Faraday on the Internet and found an article on Dr. L. Pierce Williams in Wikipedia. I downloaded the article. Dr. Williams died on February 8, 2015. When I read Arthur Zajonc's 1993 book, *Catching the Light,* he had the same information on Faraday and by his own citations in "Notes" had relied heavily on Williams biography of Faraday.

—. James Clerk Maxwell, June 13, 1831 to 1879.

Random House College Dictionary, 1999 edition. New York: Random House.

—. Zajonc, Arthur. 1993. *Catching The Light.* New York: Bantam Books.

Epilogue

Teilhard De Chardin:
The Search For The Light In Evolution

This e-Book/Audio Book began with a brief biographical sketch of Pierre Teilhard de Chardin: Priest, Scientist, Theologian, and Philosopher. He is the man who Christianized Purpose Driven Evolution.

In the prologue: Seeing, Teilhard asserted that seeing correctly is essential if we are to understand the universe as it is. In order to see correctly, we have to see ourselves, the human being, as the "center of perspective."

Teilhard maintained that because we are human, that is, have a highly-developed- consciousness with the power of reflection, we are looking at some aspect of ourselves in everything we see. The human being is the "center of perspective" and also the "center of construction of the universe," that is, the evolution of the universe (cosmogenesis) is the milieu (surroundings, environment) of which the evolution of the human phenomenon (anthropogenesis) is an integral part. Thus it is that when we look at ourselves the human beings, we grow in our knowledge of the universe.

Teilhard revealed my place as a human person in the development of the earth as Man (male and female) the most highly developed consciousness, the leading shoot of evolution, the arrow pointing the way for further evolutionary development toward Omega Point: the final fulfillment of the world through creative union with Omega (God).

He chronicled the Purpose Driven Evolution in four Parts

Part I: Prelife

At the heart of the physical world is matter, something that occupies space and has weight. But before the simple corpuscles that formed original matter, there was what Teilhard terms a kind of luminosity (light) indefinable in terms of figures.

The simple original corpuscles, consisting of protons, neutrons, electrons, photons, group together in various ways to form the simple elements like hydrogen. Complexification continues and compound bodies appear which are multicellular.

Finally, when these multicellular masses reached a certain stage of complexification, a critical stage, life appeared (*The Phenomenon of Man*, pages 47-48).

Part II: Life

When a certain stage of exterior (physical) and interior (psychic, consciousness) complexification had been reached, and conditions on earth were favorable, life was born naturally (without the contrivances of humans in a laboratory) once and once only. Although the evolutionary process was slow and gradual, there are critical stages or points before which something did *not* exist but after which it is found.

I cited the homely example of how the pressure cooker works. When the temperature within the cooker reaches a certain degree, the atoms within the molecules of whatever stuff was being cooked, move faster and faster, expanding all the while. If the cook does not watch the gauge a critical point (stage) is reached and the stuff within explodes through the opening under the gauge and one has a mess all over the stove or even on the ceiling.

Part III: Thought

As Darwin also asserted, it is precisely because the primates have relatively weak bodies that nature went to work on the brain as the instrument for their survival – and the complexity of the brain is directly related to the complexity of the within (*The Phenomenon of Man*, pages 147- 160).

When the complexification of the without (the brain and nervous system) and the within (awareness, consciousness) of a certain line of evolution (the mammals) had been reached another critical point – thought was born in one phylum the hominids – and man was born. The first man was the hominid who knew that he knew – had the power of reflection – being able to step back and examine his/her own thoughts and behavior.

With the birth of thought a new layer – the thinking layer appeared – which has spread all across the entire world because of the spherical shape of our globe. Since De Chardin's time in the early 1950s the internet has developed and in 2015 what happens in one part of our globe almost instantaneously is known everywhere. Ideas "bump" into each other and build on each other. All sorts of knowledge, especially self-knowledge, is increasing at an ever growing pace.

Part IV: Superlife (The Future)

Nearly seventy years ago Teilhard questioned whether in all our theories on evolution we have emphasized the mechanical and impersonal at the expense of the place due to "person" and the strength, the effect, the influence of personalization in the movement, and shape the evolutionary process.

Alpha (God) began the Evolutionary Process. "In the beginning was the Fire" – the Eternal Light who set everything in motion – perhaps by The Big Bang.

In Part IV Teilhard repeated his vision of the goal or purpose of evolution, that evolution is an ascent to consciousness (reflection).

This is an ascent from the simple to the increasingly complex with consciousness being the most complex interior expression of a highly complex "without" or physical structure, that is, the brain and nervous system. Since this has been the direction in which evolution has progressed to this point, and since scientific prediction for the future is based on repeated and sustained past behaviors, it seems logical to predict that evolution's fulfillment will be in some type of supreme (even more complex) consciousness (*The Phenomenon of Man*, page 258).

This supreme consciousness Teilhard calls "Omega" because this consciousness is the goal or end of evolution. Omega is personal. This is logical since personalization is simply an ever-deepening consciousness—the more highly developed the consciousness, the more we can apply the term "person" to each centered unit of such consciousness. Also, as we have seen, evolution moves from the simple to the complex, from non-life to life, from life to thought and consciousness, and thence to deepening consciousness.

Thus, if Omega is the goal of evolution, Omega must be personal, must be a supreme consciousness, a supreme person, for the personal is attributed a higher (more complex) order of existence (humankind)—and how can a higher order of existence (humankind) be capped, crowned, or have as its goal the impersonal, a lower order of existence?

Union with Omega is not a union with a kind of world spirit into which all lesser spirits will be submerged. Person is an immortal characteristic of humankind, for Teilhard does not subscribe to the concept of God as one who reabsorbs into himself the spirits of individual humans once their earth lives are ended. Omega (God, the end or goal of all things) is not a grand stew of elements that have had a temporary earthly existence and are now part of the stew. Rather, God is indeed the magnetic center whose force draws all other individual centers back to himself—to be reattached to him in an intimate, personal, eternal union, but not to a union which annihilates the individual center (person). Humans become most fully human only when they are finally grafted onto Omega, and He becomes all in all to them (*The Phenomenon of Man*, pages 262-263).

It should be noted here that Teilhard speaks of Omega point as the focus of union, while Omega is the Supreme Consciousness that is the center of the union. The relationship between Omega and individual centers of consciousness may be likened to the relationship between the planets of our solar system and their sun. The energy and attraction of the sun both gives life to its planets and holds them in an orbit about itself. Yet each of these planets remains a distinct entity—a center of existence.

Omega is in all things by the very fact that the basic stuff of the universe is psychic energy (light) and Omega is the Alpha, the source of this energy (and is Himself Psychic Energy, "The Light," centered in one Divine Nature of three Divine Persons). Alpha and Omega, the first and last letters of the Greek Alphabet, are often used to indicate that God is the beginning and end of all that exists.

I invite you now to join me as we move on to e-Book 4 in our Called into Life by the Light Series. How does *Near Death Experience: Out Of The Darkness Into The Light* reflect what happens to persons who have had a profound near-death experience that really happens to *me* when I die?